Aquatic Flora
of Northern New England

Floating, Submersed, & Floating-leaved Plants
in Maine, New Hampshire, and Vermont

Donald J. Padgett

AQUATIC FLORA OF NORTHERN NEW ENGLAND
Floating, Submersed, & Floating-leaved Plants in Maine, New Hampshire, and Vermont

Donald J. Padgett

Copyright © 2022 by Donald J. Padgett
All rights reserved

Spatterdock Press
Middleboro, Massachusetts

Printed by KDP, An Amazon.com Company

ISBN-13: 978-0-578-28465-1

Cover photograph: *Brasenia schreberi*, ©Eric McCarty, 4 February 2015

"Water is the driving force of all nature"

Leonardo da Vinci

VERMONT
MAINE
NEW
HAMPSHIRE
Aroostook
Piscataquis
Somerset
Penobscot
Washington
Hancock
Franklin
Waldo
Kennebec
Knox
Oxford
Andro-
scoggin
Lincoln
Cumberland
Sagadahoc
York
Essex
Franklin
Orleans
Grand
Isle
Lamoille
Caledonia
Chittenden
Washington
Addison
Orange
Windsor
Rutland
Bennington
Windham
Coos
Grafton
Carroll
Belknap
Sullivan
Merrimac
Strafford
Rockingham
Cheshire
Hillsboro

CONTENTS

Preface vii
Acknowledgments ix
Introduction 1

How to use this book 3
    Format   3
    Use of keys   3
    Plant names   3
    Species descriptions   3
    Illustrations & Photographs   4

Aquatic Plants of Maine, New Hampshire, & Vermont 5
    Composition   5
    Geographic distribution   6
    Features of aquatic plants   6

Keys to Genera 10
    Synopsis of Keys   10
    Submersed plants   10
    Floating & floating-leaved plants   14
    Plants with some leaves or stem portions
       extending above the water surface   16

Keys to Species & Descriptions 19

References 208
Illustration & Photograph Credits 209
Glossary 213
Index 215
About the Author 219

PREFACE

This book comes as the third in a series of guides dedicated to the aquatic plants of New England. Like that of its predecessors (*Aquatic Plants of Massachusetts*, *Aquatic Flora of Connecticut & Rhode Island*), the impetus to construct this book arose from teaching about aquatic plants and their role in assessing waterbodies and the frustration of finding an appropriate field manual for students. While there are excellent field guides on aquatic plant identification for this region, most utilize terminology far too advanced for novices or are far too simple (i.e., only including "common" species), always embrace too broad a geographic coverage which unnecessarily adds many species to the mix for consideration, and lack telling photographs. So, like its predecessors, this book was constructed as a quasi-technical—yet comprehensive—identification resource devoted to all the fully aquatic macrophytes in Maine, New Hampshire, and Vermont. It offers dichotomous keys to identify species, illustrations, color photographs, basic distribution maps, and general accounts of each species. I hope that it will be of use to all those with professional or amateur interest in the aquatic plants of this region and surrounding areas.

DJP

ACKNOWLEDGEMENTS

I am always grateful to those who helped advance my interests in aquatic plants, particularly Garrett Crow, Barre Hellquist, Donald Les, Tom Philbrick, and Jack Holt. I extend gratitude to those who granted me use of their photographs or figures, especially Katy Chayka and Peter Dziuk of *Minnesota Wildflowers* (www.minnesotawildflowers.info), Barre Hellquist, and Michael Graziano. As always, I am thankful to Bridgewater State University for continued support. Lastly, a note of love to Shannon, Morgan, Sullivan, and my pups Ali and Whiskey.

x

INTRODUCTION

Aquatic Flora of Northern New England summarizes the distribution, habitat, and biology of plants that grow in fresh, brackish, or saline waters of Maine, New Hampshire, and Vermont. It is intended to foster an appreciation of this fascinating group of plants and assist users in identifying them to species. This guide aims to update and bridge the gap between the more technical floras and the overly simple field guides of our region, with a focus on the floating, submersed, and floating-leaved flora. Consequently, advanced botanical terminology is used at a minimum, with simpler descriptive words or phrases used as a substitute.

I included all truly aquatic plant species (of angiosperms, ferns, and fern-allies) considered native or introduced in Massachusetts that one may encounter in ponds, lakes, pools, rivers, streams, and marshes. Truly aquatic (or "fully" aquatic) species are defined here as those with most of their cycle spent floating on, or submerged in, fresh, brackish, or saline water. Native or indigenous plants are those which occur naturally in Massachusetts. Introduced species entered the flora from other geographic regions either by intentional or accidental means. To make the book as comprehensive as possible, both common and rare species are included. While less likely to be encountered in the field, the rare and endangered species were combined with the more common ones so that their place in the region's rich aquatic diversity may be fully appreciated, and with the hope that new occurrences may be discovered and protected.

While 141 species are described in detail, 20 additional species are discussed for comparison purposes or represented in various ways.

What This Book Does Not Cover

While this guide covers the 141 "true" aquatic plants of the region, there are still many plants (or plant-like relatives) which occur in or around aquatic habitats that are not included in this book. Notably, this guide excludes the many "emergent" species—unless such species can also be found as a distinctive submersed growth form (e.g., *Gratiola aurea*). Emergent aquatic or wetland plants grow in shallow or damp areas typically along the edges of marshes, ponds, lakes, and rivers. Unlike true aquatic vegetation, emergent plants are rooted in the ground with the majority their stems, leaves, and flowers held above the water surface [e.g., Cattails (*Typha*), Wild rice (*Zizania*), sedges (*Carex*), etc.]. In cases where genera also contain common emergent species in the region (e.g., *Glyceria, Sagittaria, Sparganium*), the emergent species are noted in the descriptive account of the genus.

For the sake of brevity, aquatic bryophytes (non-vascular plants) are excluded too, including floating liverworts (e.g., *Riccia, Ricciocarpus*) and submersed mosses (e.g., *Sphagnum, Fontinalis*, etc.). Likewise, the few submersed macrophytic algae (i.e., *Chara, Nitella*), typical of our hard waters, are omitted. Lastly, any aquatic plant that is known to be an interspecific hybrid (e.g., hybrids in *Potamogeton* and *Isoëtes*) is not detailed here outside of noting within the parental species accounts a hybrid is known.

HOW TO USE THIS BOOK

Format. The book is arranged into an easy-to-use format. If the common name (e.g., Eurasian water-milfoil) of a plant is known, a quick check in the index can direct you to the page number for the species description. If the scientific generic (genus) name is known (e.g., *Myriophyllum*), flip through the descriptive text pages to locate the page for that plant(s). The descriptive pages for plants are arranged alphabetically by genus name, and then by species within the genus. Another method would be to thumb through the book and look for an illustration or photo of a plant like the one you are trying to identify. If the plant in question is completely unknown to you, or if you are not completely certain of its identity using the preceding methods, the plant can be identified using the taxonomic keys in the book.

Use of keys. Identification keys were constructed to facilitate the identification of species. They consist of a series of numbered couplets, with a choice of two contrasting statements. After a choice is made, a number at the end of the statement directs you to either to the next couplet or a plant name. Two sets of keys are provided to help sort out this large group of plants: 1) keys to genera, and 2) keys to species within a genus. Once a genus is identified in the key, locate the genus descriptive page. If there are more than one species in the genus, each descriptive entry begins with an overview of the group and ends with a key to the various species.

Plant names. The scientific plant names in this book largely agree with *Flora Novae Angliae* (Haines 2011). To make the book useful to a wide audience, the plants are referred to by common names in addition to their scientific names (except for keys). If a species has more than one popular common name, alternate common names are provided within the text.

Species descriptions. At the top of each species descriptive page the **common name** and accepted **scientific name** and author (who named the species) is given. The family name in scientific form is included if the species is the only one for the genus and there is no genus descriptive page.

A description of the species is then provided with features useful for its identification emphasized. Descriptions are organized with a sequence of plant type, stems, leaves, flower clusters, flowers, and then fruits, and are based on the feature's appearance during the active growing season. If a feature is not especially useful for identifying a plant (e.g., anther number), it is not mentioned. **Habitat & Distribution** indicates native status, provides the common habitat types the species occupies, and its general range with an attempt to characterize its abundance in the region. If the species is listed as rare or prohibited in ME, NH, or VT this information is noted here. **Notes** provides information regarding any specific environmental conditions the plant requires or offers, its role in a community, and similarity to other plant species.

   A **map** shows the distribution of each species as documented within the counties of Maine, New Hampshire, and Vermont and includes known introductions or invasions. A county is shaded in which the species is known to occur. The maps differentiate by color between native (green) and introduced (red) occurrences.

Illustrations and photographs. Each species account is supplemented with photographs and or illustrations. Attributable credits of each illustration or photo are found on Page 209.

AQUATIC PLANTS OF ME, NH, & VT

Composition

Maine, New Hampshire, and Vermont are each water-rich states and collectively contain more than 7500 lakes, 58,000 miles of rivers, and >5275 miles of estuarine waters which host a diverse assemblage of aquatic plants. The northern position of these states places the region at the intersection of major bioregions (exerting boreal, southern, coastal influences). This convergence leads to a great aquatic biodiversity and unique collections of species in the different areas. For example, some hydrophyte species in northern New England are at the northern extent of their geographic ranges (e.g., *Nuphar advena*, *Proserpinaca pectinata*), while others are at the southern end of their ranges (e.g., *Hippuris vulgaris*, *Littorella americana*).

The 141 hydrophytes considered in this book include three (of the four) plant life forms associated with aquatic habitats comprising 130 flowering plants (69 monocotyledons, 61 dicotyledons), 9 quillworts, and 2 ferns. These "true", or strictly aquatic, hydrophytes range from plants that grow completely submerged to floating-leaved and free-floating species. Most species (87%) are native, but 13 are considered non-native. Plants that may grow with their roots and lower stems under water but hold a more substantial amount of their body in the air—so called *emergent* species—are important components of a state's flora but omitted from this guide, as with lower plants, like aquatic liverworts, mosses, and Charophyte algae. Those few emergent species (i.e., *Gratiola aurea*, *Hypericum boreale*, *Pontederia cordata*) that also possess a distinctive submersed growth form, however, are included.

While most of these aquatic plants may be commonly encountered, 50 species are classified as endangered, threatened, or of special concern by Maine's Department of Inland Fisheries and Wildlife, New Hampshire's Natural Heritage Bureau, or Vermont's Department of Fish & Wildlife. Individuals of listed species are protected under provisions of the Maine Endangered Species Act of 1975 (MRS Title 12, § 12801), New Hampshire Native Plant Protection Act of 1987 (RSA 217-A), or Vermont Endangered Species Law (10 V.S.A., Chapt. 123).

Most non-native aquatic plants introduced into Maine, New Hampshire, and/or Vermont are deemed invasive and banned from sale or distribution. Invasive aquatic plants have impacted many lakes, ponds, reservoirs (>95 lakes in VT, 76 in NH, 35 in ME) and rivers and streams (21 rivers in VT, 11 in NH, have invasives). Exotic species like *Hydrilla verticillata*, *Hydrocharis morsus-ranae*, *Myriophyllum spicatum*, *M. heterophyllum*, *Najas minor*, *Potamogeton crispus*, and *Trapa natans* are infamous weeds that threaten native species and ecosystems.

Geographic Distribution

The natural occurrence of certain hydrophytes is largely regulated by chemical properties of their watery environment, as well as water depths and substrate preferences. As with all aquatic organisms, the optimal growth and survival of hydrophytes is influenced by levels of pH, alkalinity, and interactions of elements in the water. The chemical composition of most New England surface waters shows strong regional patterns related to the underlying geology (Mattson et al. 1992; Norton et al. 1989; Thompson and Sorenson 2000). Low pH and alkalinity levels are typical in waters where bedrock consists of granite, sand, and gravels. Where the geologic substrate shifts (westward and northward) to a prevalence of limestone and marble, pH and alkalinity levels increase. This pattern is reflected, to a degree, in the geographic distribution of many aquatic plants.

The high alkalinity, pH, and calcium levels characteristic of waterbodies in western Vermont or northern Maine support species otherwise seldom found elsewhere, like *Heteranthera dubia, Myriophyllum sibiricum, Ranunculus longirostris, Sagittaria cuneata, Stuckenia filiformis,* and select species of *Potamogeton (P. friesii, P. hillii, P. richardsonii, P. strictifolius).* The limy, hard waters cause higher available nutrients, so these waterbodies tend to have lower water quality and clarity. Conversely, more highly acidic waters in high elevation mountain lakes or rain-fed lowland seepage lakes nearer to the coast—underlain by non-calcareous geologic substrates—support species like *Isoëtes echinospora, Littorella americana, Potamogeton confervoides, Proserpinaca pectinata, Sagittaria teres,* and *Subularia aquatica.* These soft waters are nutrient-poor and have high clarity.

Watercourses influenced by ocean tides, dominated by Na and Cl, support *Crassula aquatica, Eriocaulon parkeri, Lilaeopsis chinensis, Limosella australis, Sagittaria montevidensis,* and *Zannichellia palustris* in brackish conditions, and *Ruppia maritima* and *Zostera marina* in more saline conditions. Of course, numerous hydrophytes are more broadly tolerant of water chemistry. This is reflected in their more widespread distributions (e.g., *Elodea nuttallii, Nuphar variegata, Nymphaea odorata, Potamogeton berchtoldii, P. epihydrus, P. spirillus, P. natans, P. amplifolius, Utricularia vulgaris,* and *U. purpurea*).

Features of Aquatic Plants

The ability to distinguish among the many kinds of aquatic plants is based foremost on a knowledge of their forms and characters. Since hydrophytes are derived from terrestrial ancestors, this group of species has developed shared features to overcome the challenges presented with life in water. It is these characteristics that we often rely on in identifying a species.

Keep in mind aquatic plants are notoriously variable in form. Local variations in water depth or flow can influence leaf size, shape, and texture, or presence of flowers or fruits which can make identification difficult—if even possible—at times.

<u>Morphology</u>. True aquatic macrophytes can be submersed (plant anchored to substrate or suspended), floating-leaved (plant anchored to a substrate), or free-floating on water surface (unattached to a substrate). Leaves growing submersed under water are often thin and delicate as to offer little resistance to potentially destructive water actions and to permit more light to penetrate tissues. Likewise, submersed foliage is commonly ribbon-like (fig. 1a) or highly divided (fig. 1c,d) in

form. Submersed stems, like leaves, are often thin and supported by water, lacking strengthening agents in their tissues. Consequently, submersed stems and leaves are generally limp and collapse when held out of water. Stems and leaves of certain species are specialized to be buoyant. These organs have air trapped in swollen tissues to reach light closer to the water surface or aid in keeping flowers above water. Floating leaf blades are usually thicker, at times leathery, and capable of shedding water (fig. 1b). Some species possess both floating and submersed leaves on the same plant (fig. 1c).

Reproduction. As their terrestrial relatives do, aquatic plants rely on sexual reproduction via flowers (or spores outside of the angiosperms). In the majority of hydrophytes, flowers are held above water, by various means, to utilize insect pollinators or wind. The insect-pollinated flowers are often showy, colorful (yet white in many species), and sometimes fragrant. In contrast, some hydrophytes develop flowers below water and employ very unusual underwater pollination mechanisms.

Aquatic plants are noted for their ability to reproduce and spread, often aggressively, by asexual means through rapid growth, fragmentation, and or formation of dormant structures. Due to this propensity, hydrophytes are some of our most troublesome weeds (e.g., *Myriophyllum spicatum*, *Hydrilla*). Specialized overwintering, vegetative organs called turions, or "winterbuds", are prevalent among aquatic plants. These dormant, condensed shoots are found mostly in submersed and free-floating species. They are particularly common in species of *Potamogeton*, *Myriophyllum*, and *Utricularia*, and often useful for identification purposes.

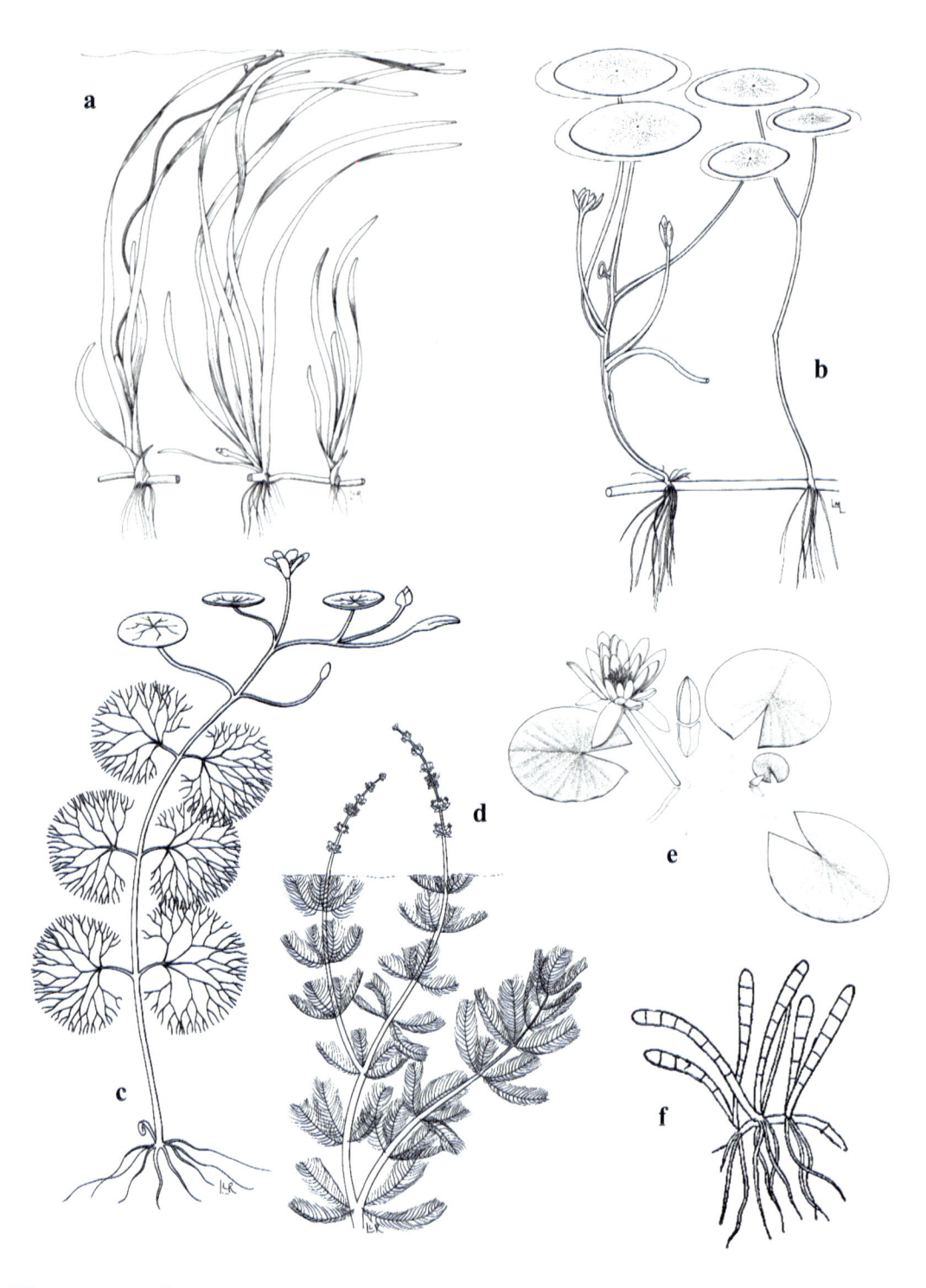

Figure 1. a. *Vallisneria*, with submersed leaves; b. *Brasenia*, with floating leaves; c. *Cabomba*, with submersed and floating leaves; d. *Myriophyllum*, submersed plant with emersed spike; e. *Nymphaea*, floating leaves and emersed flower; f. *Lilaeopsis*, submersed leaves on horizontal stems in substrate

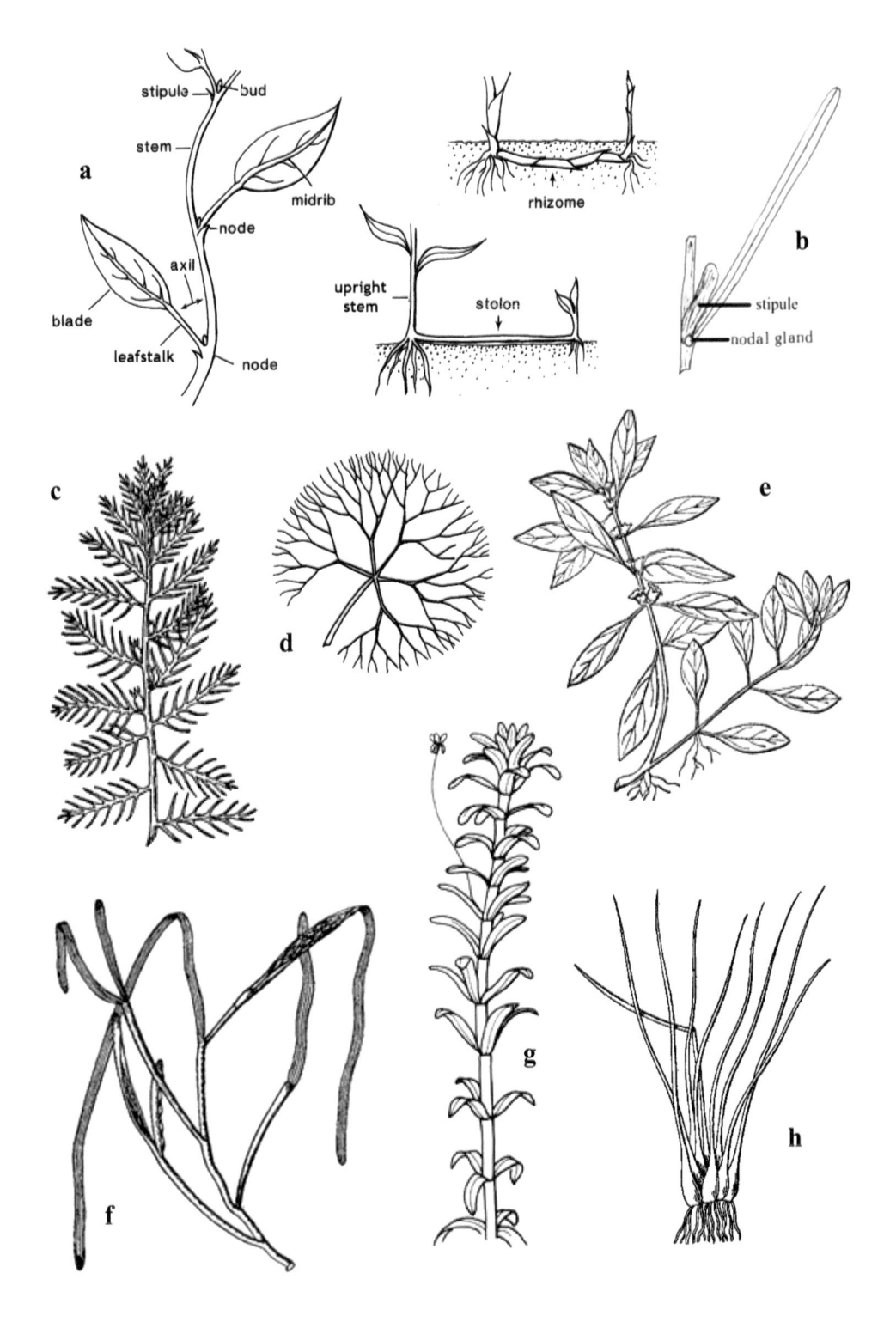

Figure 2. a. Generalized plant morphology; b. *Potamogeton*, node; c. *Proserpinaca*, pinnately divided leaves; d. *Cabomba*, palmately divided leaf; e. *Ludwigia*, undivided, opposite leaves; f. *Zostera*, ribbon-like, alternate leaves; g. *Elodea*, whorled leaves; h. *Isoëtes*, basal leaves.

KEYS TO GENERA

Synopsis of Keys

1. Plants with stems and leaves entirely submersed, normally growing completely beneath the water's surface (may be temporarily exposed to air with receding water or low tides); only flowering structures may sometimes be held above water surface (fig. 1a,d)............**Key I**
1. Plants with stems or leaves exposed to air either by floating on, or held above, water surface; flowering structures may or may not also be held above water surface...............2
    2. Plants with at least some leaves floating on the water surface (fig. 1b,c).............**Key II**
    2. Plants with some leaves or stem portions extending above the water surface.......**Key III**

Key I: Submersed plants

1. Leaves or leaf-like stems scattered along an upright stem (fig. 1c,d)................................2
1. Leaves or leaf-like stems attached basally (figs. 1a, 2h) or grouped on horizontal stems in substrate (fig. 1f) 39
    2. Leaves or leaf-like branches divided into multiple narrow segments (figs. 1d, 2c,d)........ 3
    2. Leaf blade undivided (fig. 2e-g) 19
3. Leaves or leaf-like branches whorled along stem (figs. 1d, 2g) 4
3. Leaves opposite or alternate along stem (figs. 1c, 2e,f)..9
    4. Leaves of complicated structure, each flattened leafstalk tipped with 4–8, fine-toothed, bristles (6–8 mm long) surrounding an orbicular blade (2–3 mm long) consisting of two semicircular halves, hinged along the midrib...*Aldrovanda*
    4. Leaves without bristles and a hinged orbicular blade at tip of leafstalk......................5
5. Leaf-like branches bearing small bladders at tips; flowers purple......................*Utricularia*
5. Plants lacking bladders at tips; flowers not purple.. 6
    6. Leaf segments serrated with small teeth, leaves somewhat stiff; plants rootless..*Ceratophyllum*
    6. Leaf segments without teeth, leaves soft in texture; plants rooted.........................7
7. Submersed leaves repeatedly divided from base; flowers borne in a conspicuous, yellow, daisy-like cluster... *Bidens*
7. Submersed leaves feather-like, with a central axis; flowers inconspicuous 8
    8. Leaf segments flattened; leaves (2–13 cm long) crowded at top of stem; flowers white, emersed on tall (to 40 cm), sometimes inflated, stems in spring/early summer.... *Hottonia*
    8. Leaf segments thread-like; leaves (1–4.5 cm long) scattered along most of stem length; flowers reduced, not white, either submersed in leaf axils or emersed on short spikes, throughout most of season... *Myriophyllum*
9. Leaves opposite...10
9. Leaves alternate..12

10. Submersed leaves pinnately divided, with a central axis; upper portion of stem often with a floating cluster of triangular to diamond-shaped blades and stout, spongy stalks...*Trapa*

10. Submersed leaves palmately divided (repeatedly divided from base); lacking a floating cluster of triangular to diamond-shaped blades... 11

11. Leaves with a distinct stalk (1–4 cm long); flowers white; a few inconspicuous, narrow (1–4 mm wide) floating leaves also produced at tips of stems during periods of flowering only...*Cabomba*

11. Leaves stalkless (crowded and appearing whorled); flowers yellow, in a daisy-like cluster; emersed, undivided toothed leaves sometimes present..................................... *Bidens*

12. Plants with leathery stems and creeping, green roots adhering to ledge, rocks, or boulders with fleshy disks; plants of rocky, fast-flowing rivers and streams.... *Podostemum*

12. Plants lacking fleshy, adhesive disks; plants of lakes, ponds, or sluggish rivers and streams... 13

13. Leaves without one central axis (midrib)... 14

13. Submersed leaves each with a central axis.. 15

14. Plants with small sac-like bladders, mostly scattered on sides of leaf-like segments; plants rootless, mostly free-floating or anchored by buried branches; flowers yellow...*Utricularia*

14. Plants lacking bladders; plants rooted to substrate; flowers white or yellow..*Ranunculus*

15. Plants terminated with a floating rosette (to 50 cm diam.) of broad, triangular to diamond-shaped leaves; fruit large (2–4.5 cm wide), hard, black, and spiny................................... *Trapa*

15. Plants without a floating rosette of broad leaves; fruit smaller and without spines.........16

16. Leaf divisions of submersed leaves flattened..17

16. Leaf divisions of submersed leaves thread-like (sometimes also flattened in *Rorippa*)...18

17. Leaves (2–13 cm long) becoming crowded at top of stem near water surface; flowers with 5 white petals, whorled on long (to 40 cm), emersed (perhaps inflated), flowering stems in spring/early summer.. *Hottonia*

17. Leaves (1–4 cm long) spaced evenly along most of stem length; flowers without petals, borne in leaf axils of emergent leaves... *Proserpinaca*

18. Leaves alternate, or both alternate and whorled on same plant, crowded along stems; flowers reduced, borne in axils of submersed leaves or arranged on emersed tight spikes .... *Myriophyllum*

18. Leaves strictly alternate, not crowded along stems; flowers conspicuous with 4 white petals, borne in emersed branching clusters... *Rorippa*

19. Leaves whorled along stem..20

19. Leaves opposite or alternate..23

20. Leaves 6–12 per whorl; stems unbranched, upper stem segments may be emergent (with fewer, shorter leaves); flowers inconspicuous, borne above water on emersed portions of stems... *Hippuris*

20. Leaves 3–8 per whorl; stems branched, completely submersed; flowers white, borne at water surface... 21

21. Leaves in whorls of 4–8.. 22

21. Leaves in whorls of 3 (some leaves opposite in pairs) .... *Elodea*

22. Leaves mostly in whorls of 4, blades 15–40 mm long, margins lacking conspicuous teeth .... *Egeria*

22. Leaves in whorls of 4–8, blades 8–15 mm long, margins conspicuously saw-toothed... *Hydrilla*

23. Leaves opposite..24

23. Leaves alternate.. 31
  24. Submersed leaves linear 25
  24. Submersed leaves lance-shaped, egg-shaped, or rounded..................................... 27
25. Leaf margins with minute teeth (visible with hand lens), blades widened at base and sheathing stem... *Najas*
25. Leaf margins without teeth; blades not widened at base 26
  26. Submersed leaves (>1 mm wide) notched at tip; stipules lacking; plants usually with a cluster of floating egg-shaped leaves near stem tip; fruit 2-lobed.................. *Callitriche*
  26. Submersed leaves (≤1 mm wide) pointed at tips; stem sheathing stipules at base of leaves; plants without floating leaves; fruit not lobed, slightly toothed on outer edge.........
  .... *Zannichellia*
27. Plants diminutive (usually <5 cm tall) and mat-forming..28
27. Plants larger (usually >5 cm tall) and not mat-forming..30
  28. Plants somewhat fleshy and stout; leaves lance-shaped and pointed at tips.............29
  28. Plants not fleshy; leaves lance-shaped, egg-shaped, or rounded, and rounded, pointed, or notched at tips.. *Elatine*
29. Bases of opposite leaves fused to form a boat-shaped cup at each node; flowers borne singly in leaf axil.. *Crassula*
29. Bases of leaves not forming a boat-shaped cup; plants sterile........................ *Gratiola*
  30. Leaves small (6 mm long), rounded, and without stalks; stems flaccid and mostly unbranched; plants without flowers.. *Hypericum*
  30. Leaves larger (10–50 mm long), oval to egg-shaped, and conspicuously stalked; stems many-branched and reddish; flowers borne singly in leaf axils..................... *Ludwigia*
31. Leaves inconspicuous and reduced to minute, stubby scales (<1 mm long); stems mostly unbranched arising singly from buried rhizomes.. *Myriophyllum*
31. Leaves conspicuous (>3 cm long); stems branched...32
  32. Submersed leaves linear, ribbon-like or thread-like, 0.2–12 mm wide, with or without an evident mid-vein, blades lacking a stalk; sheathing stipules present or absent at leaf base...33
  32. Submersed leaves elliptic, oblong, or lance-shaped, 3–75 mm wide, with an evident mid-vein, blades with or without a stalk; sheathing stipules present at leaf base...............
  .... *Potamogeton*
33. Submersed leaves flat and ribbon-like, 1–12 mm wide; stipules may or may not be present at leaf base.. 34
33. Submersed leaves thread-like <1 mm wide; stipules present at leaf base......................37
  34. Leaves with an evident mid-vein; stipules present (2–100 mm long); plants may also have stalked floating blades; flowers and fruits borne in cylindrical spikes...................
  .... *Potamogeton*
  34. Leaves without a mid-vein; stipules lacking; floating blades absent; flowers and fruits otherwise..35
35. Leaves to 10 cm long, 1–5 mm wide; flowers yellow with six petal-like lobes, borne singly; fruit an elongated capsule.. *Heteranthera*
35. Leaves 30–100+ cm long, 2–12 mm wide; flowers not yellow, borne in clusters; fruit a beaked drupe.. 36
  36. Leaves not sheathing the stem; flowers borne at or above water surface in dense, spherical clusters; plants of freshwaters .... *Sparganium*
  36. Leaves sheathing the stem; flowers submersed, enclosed within a leaf-like sheath; plants strictly marine.. *Zostera*
37. Leaves 1-veined and blunt-tipped; floating blades absent; flowers and fruits borne underwater; fruits in clusters of four, each fruit on a long stalk (1–2 cm long); plants chiefly

coastal in saltwater, with thread-like rhizomes.. *Ruppia*
37. Submersed leaves with 1–5 veins and sharp or blunt tips; floating blades sometimes present; flowers and fruits borne on crowded, cylindrical spikes mostly above water; plants of freshwaters, mostly lacking rhizomes.. 38
    38. Leaves stiff, channeled, 1-veined; floating blades absent; stipules fused to leaf base for more than two-thirds of stipule length; rhizomes bearing small (1–2 cm), white tubers.. *Stuckenia*
    38. Leaves flaccid, flat, with 1–5 veins; floating blades sometimes present; stipules free, or if fused to leaf base, then for less than half of stipule length; rhizomes lacking (or if present, not bearing tubers) .... *Potamogeton*
39. Leaves quill-like (linear and round in cross-section) and clustered into basal rosettes......40
39. Leaves (or leaf-like stems) thread-like, flattened, and/or more ribbon-like, clustered into basal rosettes or not.. 44
    40. Roots with cross walls; flowers, if present, to 15 mm wide with 3 white petals.. *Sagittaria*
    40. Roots without cross walls; flowers, if present, small (3–4 mm wide) with 4–5 white petals.. 41
41. Horizontal stems present at or near surface connecting plants; leaves widest in middle; flowers, if present, each with 4–5 petals, and borne singly on stalks from plant base........ 42
41. Horizontal stems absent; leaves widest at base; flowers, if present, each with 4 petals and borne in erect clusters.. 43
    42. Leaves fleshy; flowers with 5 petals; plants chiefly coastal of tidal waters......... *Limosella*
    42. Leaves not fleshy; flowers with 4 petals (flowering/fruiting only if plants emersed from water); plants of lake waters.. *Littorella*
43. Flowers/fruits not produced; leaves mostly >5 cm long, with four hollow lengthwise chambers, with a few cross-walls, swollen at bases.. *Isoëtes*
43. Flowers with 4 petals and/or fruits clustered on an erect stem; leaves mostly <4 cm long, without cross-walls.. *Subularia*
    44. Leaves (or leaf-like stems) thread-like, attached singly or grouped in clusters along horizontal stems 45
    44. Leaves flattened and/or more ribbon-like, clustered into basal rosettes 48
45. Leaves fleshy, 1–2 mm wide; flowers white.. *Limosella*
45. Leaves or leaf-like stems thread-like (<0.5 mm wide) and delicate; flowers yellow, purple, or brown.. 46
    46. Horizontal stems thick (3–4 cm diameter) .... *Juncus*
    46. Horizontal stems, if present, thread-like (<2 mm diameter) 47
47. Flowers showy, yellow or purple; leaves completely or mostly hidden in substrate; plants only apparent when flowering; a few tiny (0.2–1.1 mm long) sac-like bladders may be present, also buried in substrate.. *Utricularia*
47. Flowers not showy, borne in small (2–8 mm long) clusters behind overlapping brown scales; leaf-like stems upright, not hidden in substrate; plants apparent with or without flowers; bladders not produced.. *Eleocharis*
    48. Leaves thin, flaccid, and ribbon-like.. 49
    48. Leaves flattened but thicker, not ribbon-like and flaccid.. 51
49. Leaf blade with a 3-zone appearance (light-colored middle zone bordered by darker zone on sides); flowers small (2 mm wide) floating at surface with transparent petals; fruit a cylindrical capsule borne at end of a long, corkscrew-like stalk.. *Vallisneria*
49. Leaf blade without a 3-zone appearance; plants often sterile, but, if flowers present, larger and emersed above water; if fruit present, not borne on a long, corkscrew-like stalk.......50
    50. Roots with cross walls; leaves with cross veins running perpendicular to central mid-

      vein, blades often long (>25 cm)... *Sagittaria*
    50. Roots without cross walls; leaves without cross veins, blades 6–25 cm long...*Pontederia*
51. Plants low and small, leaves mostly 1–6 cm long...52
51. Plants larger, leaves to 35 cm long, sometimes thick and spongy.................... *Sagittaria*
    52. Leaves hollow, widest above, and blunt at tips; roots without conspicuous cross walls...
   53
    52. Leaves not hollow, widest at base and tapered to a point; roots with conspicuous cross walls... *Eriocaulon*
53. Leaves numerous, each with two hollow chambers in cross section; flowers conspicuous, clustered on long (>20 cm long) stalks; fruit a nodding capsule.......................... *Lobelia*
53. Leaves 1–few, each hollow but transversely divided with 4–6 cross walls; flowers inconspicuous, and borne on stalks about as long as the leaves; fruit a nutlet with 5 corky ribs... *Lilaeopsis*

KEY II: FLOATING & FLOATING-LEAVED PLANTS

1. Whole plant is free-floating or suspended at water surface; plants either lacking roots or with roots extending into water column..2
1. Only the leaves or branches are normally floating at water surface; plants typically rooted in the substrate (fig. 1b, e) 6
    2. Plants diminutive, ≤1.5 cm long; if present, roots unbranched; no recognizable flowers present...3
    2. Plants larger, leaves 5–40 cm long and spongy; roots branched and feathery; flowers large (5–7 cm wide), showy, and blue or purplish-blue......................... *Eichhornia*
3. Leaves distinguished from stems; leaf blades fuzzy and positioned in two rows along horizontal stems...*Azolla*
3. Plant body not differentiated into distinct stems and leaves; plants not fuzzy................4
    4. Plants with roots below, extending into water; bodies flattened throughout, 1–4 mm wide..5
    4. Plants without roots; bodies spherical, oval, or boat-shaped (i.e., flattened above and rounded below), 0.5–1 mm wide... *Wolffia*
5. Plants each with one root below; bodies green beneath............................*Lemna*
5. Plants each with 7–21 roots below; bodies reddish purple beneath .... *Spirodela*
    6. Plant with a floating rosette of noticeably inflated (swollen and spongy) leaf stalks or branches; submersed leaves finely divided with thread-like segments......................7
    6. Plant lacking a floating rosette of inflated branches or leaf stalks; submersed leaves, if present, not finely divided..9
7. Floating leaves or upright branches green; bladders absent on plant; flowers white............8
7. Wheel-like rosette of whitish branches floating flat on water surface, functioning as a float for the flowering stem; numerous bladders present on submersed portion of plant; flowers yellow... *Utricularia*
    8. Rosettes of floating leaves, with large-toothed, triangular to diamond-shaped, flat blades; flowers conspicuous; fruit large (20–45 mm wide) with stout spines............*Trapa*
    8. Rosettes of erect stalks, bearing well-spaced whorls of inconspicuous flowers; fruit small (1.5–3 mm wide) lacking spines.. *Hottonia*
9. Floating leaves linear and ribbon-like, > 10 times as long as wide, without obvious stalks, usually just tips of leaves floating...10

9. Floating leaves with broad blades, not ribbon-like, with obvious leaf stalks; usually the entire blade is floating……………………………………………………………………………13
    10. Leaves 10–15 mm wide, with conspicuous cross veins (short veins running perpendicular to longer veins); flowers borne either below water or on floating stems….……………11
    10. Leaves 2–10 mm wide, without cross veins; flowers borne above water on emergent stems……………………………………………………………..……………………......…12
11. Leaf with a 3-zone appearance (a light-colored middle zone bordered by a darker zone on each side), blade not widening at tip; flowers with transparent petals; fruit cylindrical (5–12 cm long), borne singly at end of a coiled stalk; roots without cross walls………...…*Vallisneria*
11. Leaf without a 3-zone appearance, blade sometimes widening at tip; flowers showy, each with three white petals, borne in clusters at end of a long, usually floating, stem; fruit flattened and seed-like (to 2.5 mm long), crowded into spherical clusters; roots with cross walls…..………………………………………………………………………………. *Sagittaria*
    12. Leaves 15–45 cm long, dark bluish-green, with a prominent central vein; base of leaf forming a tube-like sheath around the stem; flowers and fruits in minute, narrow (1–2.5 mm wide), soft clusters………………………………………………………………*Glyceria*
    12. Leaves 50–150 cm long, green, lacking a central vein; base of leaf not tubular; flowers in spherical (5–25 mm wide) clusters developing into hard, bur-like fruit clusters …………… ……………………………..…………………………………………………. *Sparganium*
13. Leaf stalks centrally attached to underside of floating leaf blade……………………...…14
13. Leaf stalks attached at the edge of floating leaf blade……………………………….…… 15
    14. Leaf blades are broadly oval (4–13 cm long); submerged portions of plant covered in a clear, jelly-like coating; stems flexible and buoyant; flowers red purple, 2 cm wide…..…… …..………………………………………………………………………………….*Brasenia*
    14. Leaf blades are circular, massive (to 60 cm wide); submerged portions of plant not covered in a jelly-like coating; stems creeping on substrate; flowers yellow, 10–20 cm wide…..…………………………………………………………………………… *Nelumbo*
15. Leaves are all basal, arising from base of plant close to substrate……………………..…………16
15. Leaves attached along an upright stem in the water column …………………………….21
    16. Floating leaves divided into segments or deeply lobed along sides………………….…17
    16. Floating leaves not divided into segments or lobed along sides (blades may have a basal sinus) ………………………………………………………………………………….. 18
17. Floating leaf comprised of 4 triangular segments (each 7–21 mm long, 6–19 mm wide), resembling a four-leaf clover; reproductive structures (not flowers) submersed, hairy, pea-shaped (5 mm), near base of leaf stalks………..……….…………………………*Marsilea*
17. Floating leaf simple, blade deeply lobed and divided into 3–5 main segments; flowers emersed above water on branching clusters, each with five, small (2–5 mm long), yellow petals……..…………………………………………………………………. *Ranunculus*
    18. Plants also with submersed, linear, phyllodial leaves (25–450 mm long, 15–110 mm wide); roots with cross walls; fruits flattened and seed-like (to 2.5 mm wide), crowded into spherical clusters………………………………………………………………*Sagittaria*
    18. Plants lacking submersed, linear, phyllodial leaves; fruits globose, berry-like (4–40 mm wide) and borne singly; roots without cross walls……………………………………….19
19. Leaves large (mostly >8 cm wide), veins branched and running toward margins; flowers bisexual, with >3 white, yellow, or pinkish petals; fruits large (4–40 cm wide)…..……….20
19. Leaves smaller (mostly >6 cm wide), veins unbranched and running toward tip; flowers unisexual (on different plants) with 3 whitish petals; fruits smaller (<1 cm wide)…………… ………………………………………………………………………………….*Hydrocharis*

20. Leaf blade generally circular in shape, with pointed lobes, and usually purplish on underside; leaf stalk round in cross section; flowers white to pink, 3–19 cm wide with numerous petals; fruits maturing underwater on coiled stalks………………………*Nymphaea*
20. Leaf blade circular-oblong, with rounded lobes, and green on underside; leaf stalk round or flattened; flowers bright yellow, 1–4.5 cm wide, with 5–6, petal-like sepals; fruits maturing at water surface on straight stalks………………………..…………*Nuphar*
21. Floating leaves generally round or heart-shaped with a deep sinus at base, with veins centrally radiating and branching; submersed clusters of slender fleshy roots may be present at the leaf base; flowers are emersed, showy, 5–25 mm wide, yellow or white……………………………………………………………………………………….. *Nymphoides*
21. Floating leaves elliptical, oblong, or egg-shaped, veins parallel or branching; clusters of fleshy roots absent at the leaf base; flowers are small (<5 mm wide), submersed or emersed, never yellow or white…...……………………………………...……………………...22
22. Floating leaves tightly clustered into a rosette at the stem tip, blades small (5–30 mm long) and widest toward the tip; submersed leaves arranged in opposite pairs and linear; stipule absent at base of each leaf; flowers borne 1–3 in submersed leaf axils….*Callitriche*
22. Floating leaves not clustered into a rosette, blades usually much larger (6–260 mm long) and widest near middle or base of blade; submersed leaves, if present, linear or broad, and alternate; stipule usually present at base of each leaf; flowers borne on emersed cylindrical spikes or submersed few-flowered spike-like clusters……………...23
23. Leaf veins parallel; floating and submersed leaves different in shape and texture; leaves never hairy; flowers greenish………..……………………………………….…..*Potamogeton*
23. Leaf veins pinnately branching; floating and submersed leaves no different; leaves sometimes hairy; flowers pink…...……………………………………………………*Persicaria*

KEY III: PLANTS WITH SOME LEAVES OR STEM PORTIONS EXTENDING ABOVE THE WATER SURFACE

1. Leaves or leaf-like branches held above water are linear, >10 times as long as wide………… 2
1. Leaves held above water with a more expanded blade, <10 times as long as wide…………. 8
   2. Plants with many submerged, flaccid, thread-like leaves or stems…...………………….3
   2. Plants without submerged, flaccid, thread-like leaves or stems……...…....……………..5
3. Plants with 1 (or maybe 2) fully developed, emergent leaves attached to the main stem, leaf with internal cross walls; submerged thread-like leaves with internal cross walls; flower cluster multi-branched………...……………………………………………………..............*Juncus*
3. Emergent leaves lacking (no leaves attached to main stem); submerged thread-like leaves or stems lacking internal cross walls; flower cluster unbranched…...……………...……………4
   4. Erect, emergent stems round in cross section, narrower than flower cluster; flower cluster appears to come from side of stem (because a 2–20 mm long leaf-like bract is attached below flower cluster, appearing like a continuation of the stem); submerged leaves with a sheathing base………………………………………………..…...*Schoenoplectus*
   4. Erect, emergent stems 3-sided in cross section, nearly as thick as the flower cluster; flower cluster attached to tip of stem; submerged leaf-like stems without a sheathing base…...………………… ………………………………………………………….. *Eleocharis*
5. Leaves clustered in a basal rosette, each essentially round in cross-section and thickest at base, generally tapering to the tip; flowers, if present, with petals……………………….. 6
5. Leaves not in a basal rosette, each generally flattened or thread-like; flowers are lacking

petals..7
    6. Base of plant swollen and bub-like, containing white spores; roots are not septate; leaves with four hollow lengthwise chambers, flexible or rigid; flowers, fruits or flowering stems never produced...*Isoëtes*
    6. Base of plant not swollen and bulb-like; roots are septate; leaves rigid; flowers with three conspicuous white petals..*Sagittaria*
7. Leaves conspicuous (>0.5 m long, 2–10 mm wide); flowers are grouped into several spherical clusters, each with a leaf attached below; fruits are a spherical aggregate of hard, beaked drupes..*Sparganium*
7. Plants delicate, grass-like, with short (5–8 cm tall) and slender (<0.5 mm wide) leaf-like stems; actual leaves scale-like and inconspicuous at stem base; flowers are grouped at the stem tip into a single cluster of overlapping scales; fruits are nut-like with bristles at the base..*Eleocharis*
    8. Leaves held above water are deeply divided or finely divided into narrow segments......9
    8. Leaves held above water are broad and flattened, not finely divided, but may have toothed margins..10
9. Emersed leaves greatly reduced and bract-like, <1 cm long..........................*Myriophyllum*
9. Emersed leaves not reduced, 1–3 cm long..*Proserpinaca*
    10. Emersed leaf margins are toothed...11
    10. Emersed leaf margins are without teeth...15
11. Leaves held at or above water are attached alternately at top of a submersed stem, each with a triangular to diamond-shaped blade, and a long leaf stalk with an inflated portion; fruit with 4 horn-like spikes...*Trapa*
11. Leaves held above water are alternate, opposite, or whorled on emersed portion of stems, each with a blade much not longer than wide, and a leaf stalk lacking an inflated portion; fruit with without 4 horn-like spikes...12
    12. Emersed leaves are opposite or whorled on a stem...13
    12. Emersed leaves are alternate on stem..14
13. Emersed leaves are opposite, 2 per node, and 2–4 cm long; flowers borne in a single, conspicuous, yellow, daisy-like cluster...*Bidens*
13. Emersed leaves are whorled, 4–6 per node, and 0.4–1.8 cm long; flowers borne in axils of leaves forming a leafy spike..*Myriophyllum*
    14. Flowers are unstalked, inconspicuous, purplish-greenish, without petals, borne in axils of emersed leaves; fruit is 3-sided, common; emersed leaves with sharp, forward-pointing teeth; submersed leaves sometimes absent......................................*Proserpinaca*
    14. Flowers are stalked (stalks 5–15 mm long), conspicuous, with 4 white petals, borne in cluster above the emersed leaves; fruit a rounded pod, rare; emersed leaves with spreading teeth, some blades may be untoothed as well...............................*Rorippa*
15. Emersed leaves opposite or whorled on stem, blade 0.5–4 cm long, stalkless or short-stalked (<3 cm long)..16
15. Emersed leaves arising from a basal rosette, blade 3–30 cm long, on long stalks (>6 cm long)..*Sagittaria*
    16. Emersed leaves whorled, crowded, 8–12 per whorl, and narrow (1–3 mm wide)..*Hippuris*
    16. Emersed leaves opposite, 2 per node, blades wider (to 20 mm wide) 17
17. Emersed leaves without stalks; stems green; flowers borne in a single, conspicuous, yellow, daisy-like cluster at stem tip; submersed leaves are finely divided into thread-like segments..*Bidens*
17. Emersed leaves with stalks (5–25 mm long); stems reddish; flowers are inconspicuous, green-brown, borne singly in leaf axils; submersed leaves resembling emersed leaves...*Ludwigia*

Keys to Species & Descriptions

Aldrovanda — Droseraceae
Waterwheel plant

WATERWHEEL PLANT — *Aldrovanda vesiculosa* L.

Waterwheel plant is a submersed, free-floating, rootless, perennial, with floating, branched stems (to 40 cm long). Leaves are 10–15 mm long, submersed, and whorled in clusters of 6–9. Each flattened leafstalk is tipped with 4–8, fine-toothed, bristles (6–8 mm long) surrounding an orbicular blade. The blade (2–3 mm long) consists of two semicircular halves, hinged along the midrib, which can rapidly close to trap invertebrates. Flowers (borne singly in leaf axils) are not produced in our area. Turions are produced at stem tips in late season.

HABITAT & DISTRIBUTION
Aldrovanda vesiculosa is a non-native, carnivorous species of stagnant lake waters. **Introduced** and uncommon in northern New England.

Notes: This unusual, globally-rare, plant is native to Africa, Asia, and Australia but was introduced into US by carnivorous plant hobbyists in 1970s. Its peculiar, fast-acting, leaf traps have one of the quickest known plant movements (trap closure takes 10–20 milliseconds). Under optimal conditions, plants are known for rapid growth.

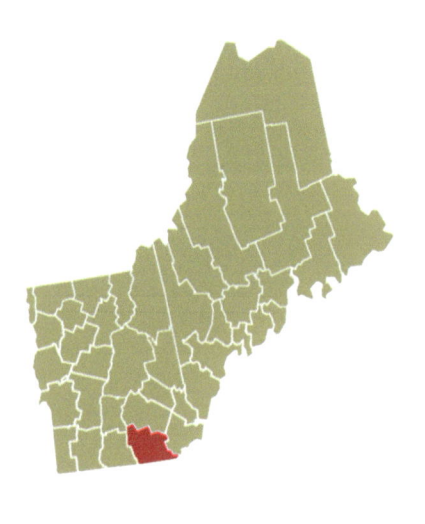

Azolla — Azollaceae
Mosquito fern

EASTERN MOSQUITO FERN — *Azolla caroliniana* Willd.

Eastern Mosquito fern is a tiny, free-floating, aquatic fern with horizontal stems (0.5–1 cm long), slender roots, and overlapping, fuzzy leaves. Leaf blades are floating, very small (0.5–2 mm wide), dark green (or reddish), unequally two-lobed, covered in hairs, and positioned in two rows along a stem. Plants reproduce by spores.

HABITAT & DISTRIBUTION

Azolla caroliniana is a non-native, moss-like, mat-forming fern species of stagnant lakes, ponds, or sluggish streams. Introduced and infrequent in ME and NH.

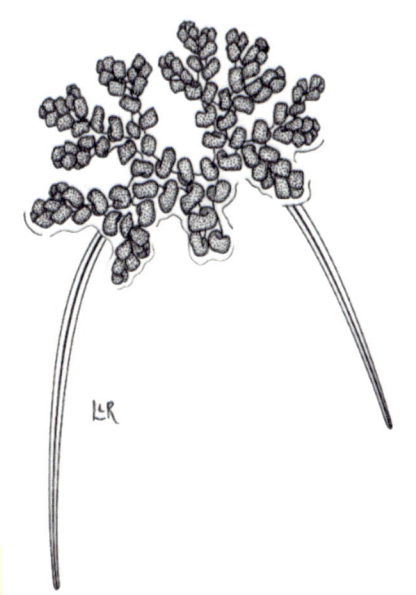

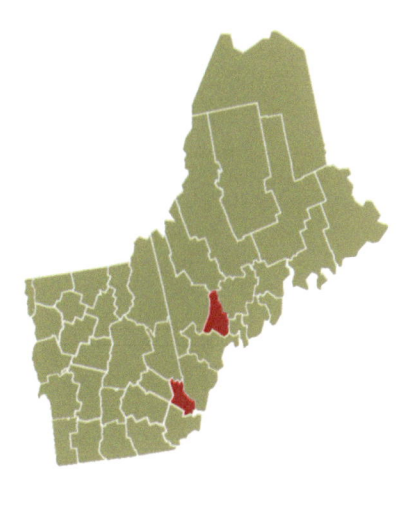

Bidens — Asteraceae
Beggar-ticks

BECK'S BEGGAR-TICKS — *Bidens beckii* Torr. ex Spreng.

Beck's beggar-ticks, or Water marigold, is a submersed, rooted, perennial with rhizomes and elongated (to 200 cm), slender, flexible stems. Leaves are opposite, stalkless, 2–4 cm long, and dimorphic: submersed leaves, appearing whorled, are finely divided into thread-like segments; and emersed leaves are simple, lance-shaped, with toothed or untoothed margins. Flowers are borne, above water, in a single, conspicuous, yellow, daisy-like cluster (head) resembling a flower (2.5 cm diam.). Fruit is a yellowish to greenish brown, linear achene (10–15 mm) with 3–6 barbed extensions (13–15 mm long) at tip.

HABITAT & DISTRIBUTION

Bidens beckii is a native species in lakes, ponds, and slow streams. It is widespread in northern New England, but rare and state-listed as *threatened* in NH. Fruits rarely mature in our region. It can occupy deep (to 7 m) waters where it will only produce submersed foliage.

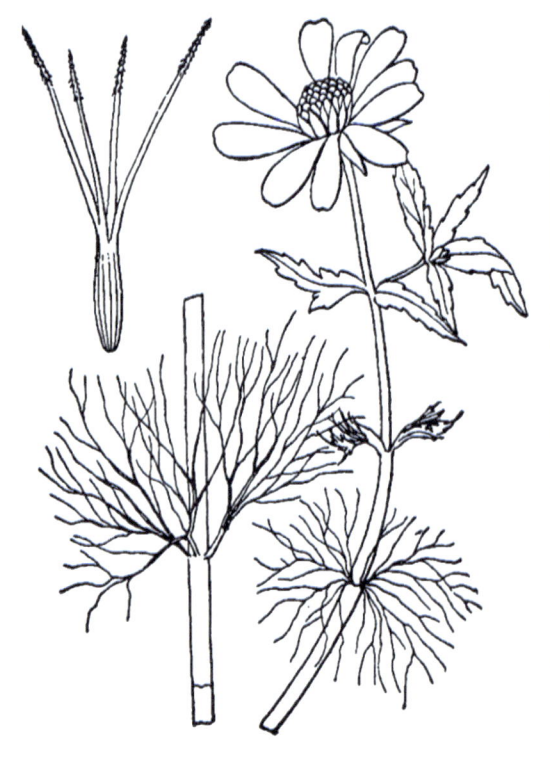

23

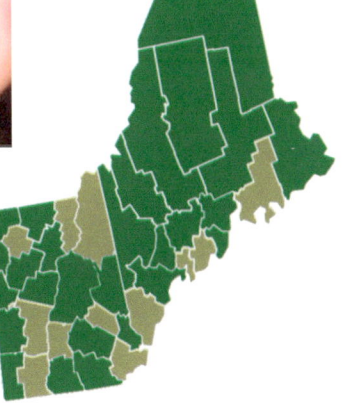

Brasenia — Cabombaceae
Watershield

WATERSHIELD *Brasenia schreberi* J.F. Gmel.

Watershield is a rooted, perennial, floating-leaved plant with creeping rhizomes and long, flexible, buoyant, branching stems. Leaves are alternate, floating, broadly oval (4–13 cm long), on long (to 60 cm) stalks centrally attached to the blades. Usually, all submerged portions are purplish and covered in a thick, translucent, jelly-like coating. Flowers (2 cm wide) are bisexual, mostly wind-pollinated, petals red purple (12–20 mm long), borne singly on elongated (to 15 cm) stalks that hold flowers about 2 cm above water. Fruit a small (6–10 mm long), leathery, few-seeded nutlet.

HABITAT & DISTRIBUTION

Brasenia schreberi is native and found in open, standing or slow waters (1–3 m deep) of lakes, ponds, and streams. It is widespread and common in northern New England. This species is highly tolerant and can reproduce asexually via rhizomes and winter buds which can result in dense, weedy colonies.

Notes: This is a very distinctive floating-leaved plant with its oval leaves and slippery, mucilaginous layer covering younger, underwater parts.

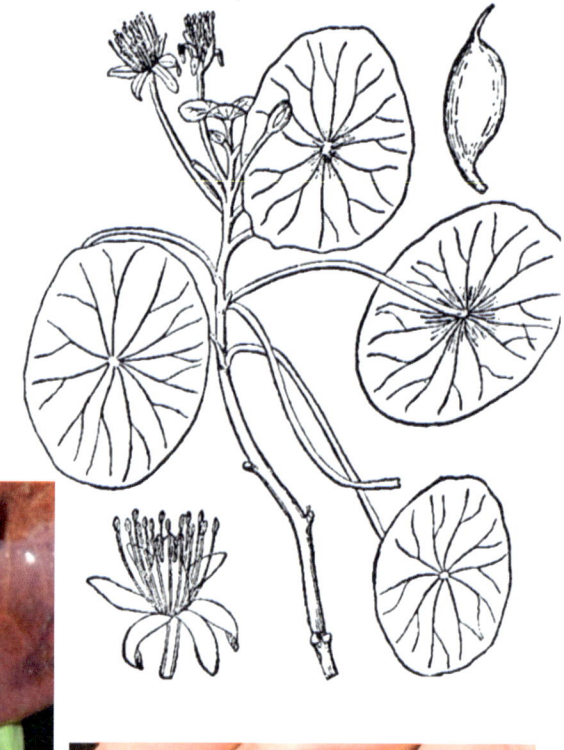

25

Cabomba — Cabombaceae
Fanwort

CAROLINA FANWORT — *Cabomba caroliniana* Gray

Carolina fanwort is a rooted, submersed species with creeping rhizomes and long (up to 2 m), slender, branching, flexuous stems. Submersed leaves are opposite, rotund (1–3.5 cm long, 2.5–5 cm wide), palmately divided (3–7 divisions) into many narrow segments, and with distinct stalk (1–4 cm long). During periods of flowering only, a few inconspicuous, narrow (5–20 mm long, 1–4 mm wide), floating leaves are also produced at tips of stems. Flowers are bisexual, small (6–15 mm diam.), white, and borne singly above water on stalks (3–10 cm) from upper leaf axils. Fruit is a small (4–7 mm long), leathery, nutlet.

HABITAT & DISTRIBUTION
Cabomba caroliniana is non-native and has become naturalized in ponds and slow streams. It is uncommon in southern NH where it is **invasive** and prohibited. This plant can become abundant and problematic as it exhibits aggressive vegetative growth, prolific stem fragmentation, and broad ecological tolerances.

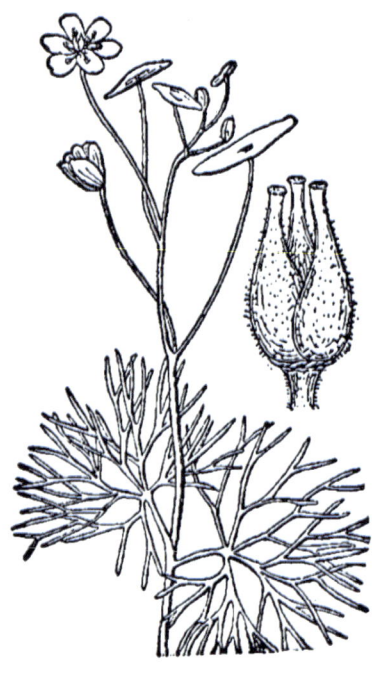

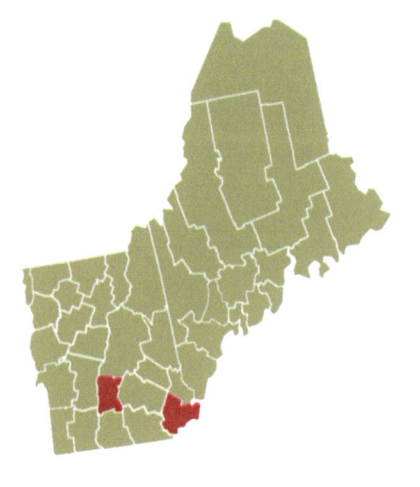

Callitriche — Plantaginaceae
Water starwort

The Water starworts are small, submerged, rooted, annuals or perennials, with slender, often flaccid, branched stems. Leaves are opposite, toothless, and mostly dimorphic: submersed leaves are linear shaped, often notched at tip, and well-spaced along the stem, and floating, broader leaves are clustered into a rosette near the stem apex. In most species (except *C. hermaphroditica*), the bases of adjacent leaves are slightly confluent, forming a narrow membranous wing across the sides of the stem (This feature is not always easily detected). Flowers are tiny, unisexual (on same plant), petal-less, and borne 1–3 per leaf axil. Female flowers have a flattened, 2-lobed ovary. Fruit splitting into four 1-seeded segments at maturity.

Four species occur in northern New England, including one non-native. The species are very variable vegetatively. Mature fruits are essential for accurate identification.

KEY TO SPECIES

1. Flowers and fruits with 2 translucent bracts (0.5–1.5 mm long) at base; leaves along the stem linear, while those crowded near the stem tip spoon-shaped, forming a floating rosette..2
1. Flowers and fruits without bracts at base; leaves all linear, never forming a floating rosette.. ....*C. hermaphroditica*
    2. Fruit 1.5–2 mm long, margins with a prominent wing extending along entire fruit length; floating leaf blades 5–8 mm wide with 5–7 veins.. *C. stagnalis*
    2. Fruit 1–1.4 mm long, margins unwinged or with a wing absent at fruit base; floating leaf blades (when present) up to 5 mm wide, with 3 (–5) veins..3
3. Fruit longer than wide by 0.2 mm or more, top portion of each segment with a thin wing, pits of fruit surface aligned in vertical rows.. *C. palustris*
3. Fruit as wide as long, lacking wings along margins, pits on fruit surface not aligned in rows... ....*C. heterophylla*

C. heterophylla

C. palustris

C. stagnalis

Autumn Water starwort — *Callitriche hermaphroditica* L.

Autumn Water starwort is a wholly submersed plant with loosely clustered stems (5–15 cm long). Leaves are all submersed, linear (4–20 mm long, 0.6–2.2 mm wide), widest at base, 1-veined, and notched at tip, and usually widely spaced along stem. Fruits rounded (1–2 mm long) with subtle but discrete winged margins.

Habitat & Distribution

Callitriche hermaphroditica is a native species of lakes, ponds, and rivers. It is rare in northern VT, where it is state-listed as *historic*. Plants are generally found in soft, nutrient-enriched waters.

Notes: Leaves tend to be darker green compared to submersed leaves of other *Callitriche* species.

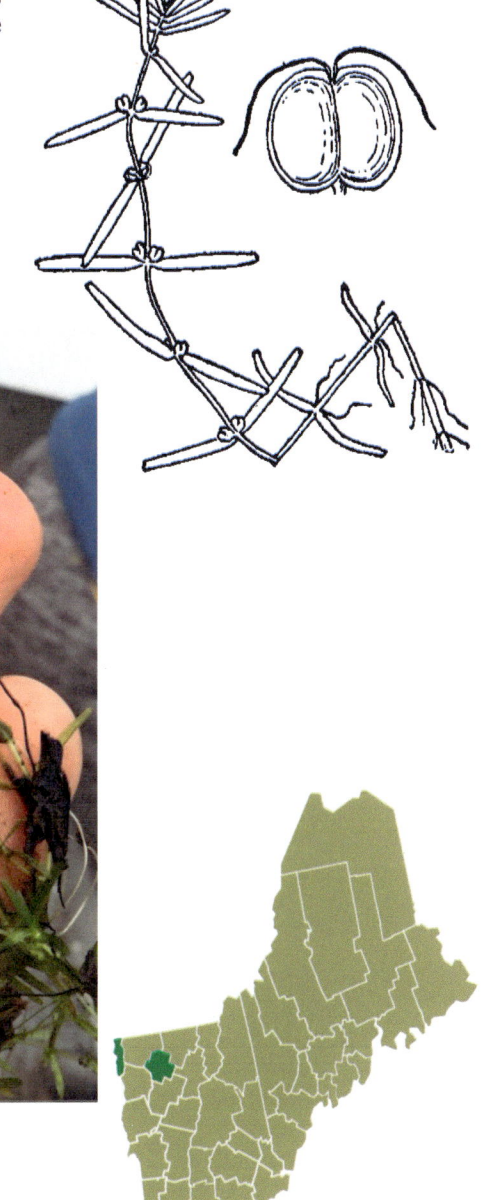

GREATER WATER STARWORT *Callitriche heterophylla* Pursh

Greater Water starwort, or Two-headed Water starwort, is a submersed plant with loosely clustered stems (10–20 cm long). Floating leaf blades (when present) are egg-shaped (5–30 mm long, 2.5–5 mm wide), widest toward the apex, blunt tipped, usually with 3 veins. Submersed leaves are linear (5–25 mm long, 1.5 mm wide) and notched at tip. Fruits about as long as wide (1–1.4 mm long, 0.6–1.2 mm wide) with unwinged margins. Pits on fruit surface irregularly distributed.

HABITAT & DISTRIBUTION

Callitriche heterophylla is a native species of lakes, ponds, bogs, mountain pools, and rivers. It is widespread but uncommon in northern New England (reaching high elevations). Rare and state-listed in ME (*special concern*). Plants are generally found in circumneutral, shallow waters.

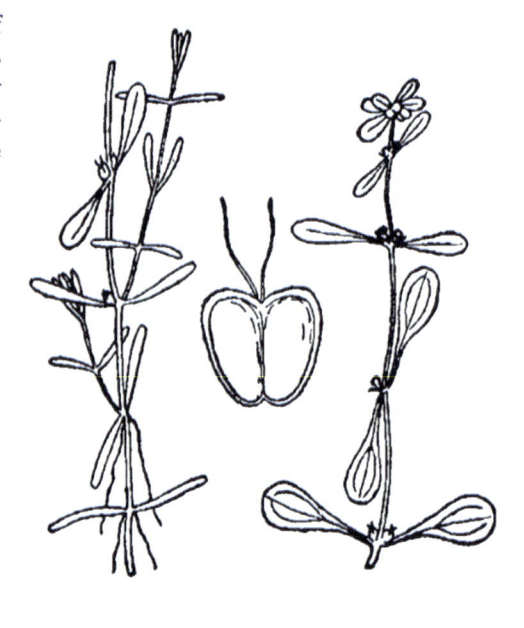

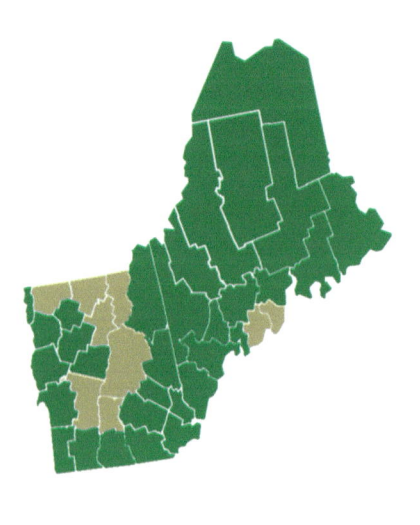

VERNAL WATER STARWORT — *Callitriche palustris* L.

Vernal Water starwort is a submersed plant with loosely clustered stems (10–20 cm long). Floating leaf blades (when present) are egg-shaped (5–30 mm long, 2.5–5 mm wide), widest toward the apex, blunt-tipped, usually with 3 veins. Submersed leaves are linear (5–15 mm long, 3 mm wide) and notched at tip. Fruits distinctly grooved, slightly longer than wide (1–1.4 mm long), widest above middle, with a thin wing along upper portions of each lobe. Pits on fruit surface aligned in vertical rows.

HABITAT & DISTRIBUTION

Callitriche palustris is a native species of springs, marshes, bogs, fens, pools, ponds, and slow rivers. It is widespread in northern New England and can reach high elevations.

Note: This species can grow as an emergent plant when exposed on wet shorelines or depressions.

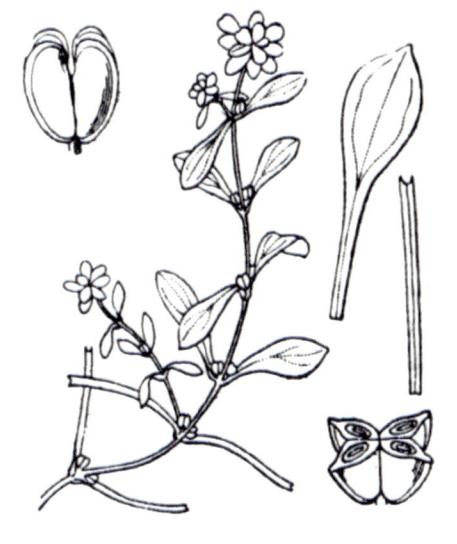

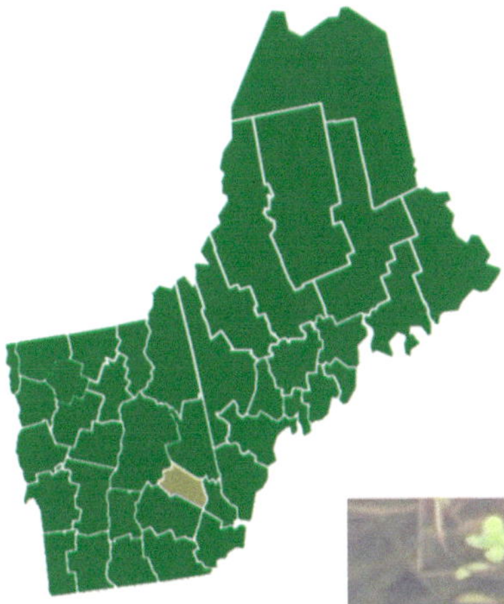

POND WATER STARWORT — *Callitriche stagnalis* Scop.

Pond Water starwort is a submersed plant with loosely clustered stems (10–20 cm long). Floating leaf blades are egg-shaped (to 20 mm long, 5–8 mm wide), widest toward the apex, blunt-tipped, usually with 5–7 veins. Submersed leaves are usually linear (4–10 mm long), 1-veined, and notched at tip. Fruits distinctly grooved, nearly round (1.5–2 mm long), with a prominent wing extending evenly along whole length of margin.

HABITAT & DISTRIBUTION

Callitriche stagnalis is a non-native species of lakes and tidal rivers. It is uncommon in northern New England. Native to Europe, the plant can spread by seed or rooted stem fragments and form dense colonies. It does not grow in highly acidic waters.

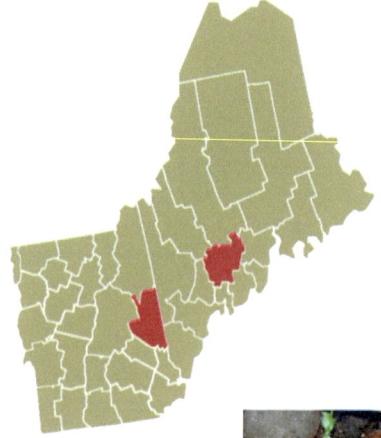

Ceratophyllum — Ceratophyllaceae
Hornwort

The Hornworts, or Coontails, are submerged, perennial, highly branched plants that float or are suspended in the water column. Plants are rootless but may become anchored to a substrate if segments become buried. Leaves are in whorls (of 3–11 leaves) along the stem, dichotomously divided, and forked into linear segments. Inconspicuous, submersed, unisexual flowers are borne in leaf axils. Fruits, if produced, are flattened, smooth-surfaced, dark green or reddish, achenes with spiny margins.

Two species occur in northern New England, both native. *Ceratophyllum demersum* is one of the most common aquatic plants in the region.

KEY TO SPECIES

1. Leaves usually forked 1–2 times and conspicuously serrated, ultimate segments somewhat flattened; fruits with 3 long spines... *C. demersum*
1. Leaves usually forked 3–4 times, with few marginal teeth, ultimate segments threadlike; fruits with 4+ long and short spines... *C. echinatum*

C. demersum

COMMON HORNWORT — *Ceratophyllum demersum* L.

Common hornwort is a highly branched, submersed plant with long (up to 3m) stems and whorled leaves. Leaves (10–30 mm long) are bright green, usually forked 1–2 times into somewhat flattened segments, curving upward, and more crowded at tips of branches. Leaf margins are conspicuously serrated with teeth, each arising from a raised green base. Flowers are small, at the leaf axils. Fruits are flattened and disk-like (3–6 mm long) with 3 long (0.5–12 mm) spines.

HABITAT & DISTRIBUTION

Ceratophyllum demersum is a native species of ditches, marshes, lakes, ponds, and slow streams. It is widespread and common throughout New England where it can become abundant. It has a relatively stiff habit with coarsely textured leaves, maintaining its shape when out of water. This makes the stems somewhat brittle whereby they can detach, disperse, and form clonal populations.

SPINELESS HORNWORT — *Ceratophyllum echinatum* A. Gray

Spineless hornwort is a highly branched submersed plant with long stems (to 1 m) and leaves in whorls. Leaves are dark green, usually forked 3–4 times into threadlike segments, and not crowded at tips of branches. Leaf margins with few, weak, inconspicuous teeth. Flowers are small, at the leaf axils. Fruits are flattened, disk-like (4.5–6 mm long), and ornamented with 3 long (1–5 mm) spines and 2–13 shorter spines between the longer spines

HABITAT & DISTRIBUTION

Ceratophyllum echinatum is a native species in lakes, ponds, bogs, and slow streams. It is widespread in New England, but rare and state-listed in ME (*special concern*). It prefers acidic, low nutrient, shaded, waters. This plant is somewhat stiff, yet not as stiff as *C. demersum*.

Crassula — Crassulaceae
Pygmy-weed

WATER PYGMY-WEED   *Crassula aquatica* (L.) Schönl.

Water pygmy-weed is a diminutive, rooted, annual species. Leaves are opposite, fleshy, linear to lance-shaped (2–6 mm long), pointed at tip, with leaf bases fused to form a boat-shaped cup at each node. Flowers are minute (to 2 mm), with 4 narrow petals, borne singly in a leaf axil. Fruit an oblong follicle.

Habitat & Distribution

Crassula aquatica is a native, tufted or mat-forming, species of open, coastal intertidal mudflats of brackish to freshwater streams, non-tidal rivers, or less commonly in ponds. It is rare and imperiled throughout northern New England, where it is state-listed in Me (*special concern*) and Nh (*endangered*).

Notes: In addition to seeds, this annual can disperse by rooted stem fragments. Interestingly, it is one of only four obligate aquatic species of an otherwise xerophytic genus of 200+ species. It is similar to plants of Waterwort (*Elatine*), but is distinguished by its fleshy leaves, fused leaf bases, and follicle fruits.

Egeria — Hydrocharitaceae
South American waterweed

BRAZILIAN WATERWEED — *Egeria densa* Planch.

Brazilian waterweed, or Brazilian elodea, is a submersed, rooted perennial with elongated (to 3 m), flexuous, branched stems. Leaves are in crowded, overlapping whorls of 4–6, lance-shaped (1.5–4 cm long, 3–6 mm wide), bright to dark green, recurved, with finely tooted margins (under magnification). Flowers are unisexual (on separate plants), white, 20 mm in diam., and borne singly from upper leaf axils on slender stalks reaching water surface. Only male plants are known in the US.

HABITAT & DISTRIBUTION

Egeria densa is a non-native species of lakes, ponds, and sluggish rivers. It is uncommon in southern NH and VT, where it is **invasive** and prohibited. The species is native to southeastern Brazil and introduced through the aquarium trade. In the US it reproduces strictly asexually via fragments, grows rapidly, and can outcompete native plants. Similar to Waterweeds (*Elodea*).

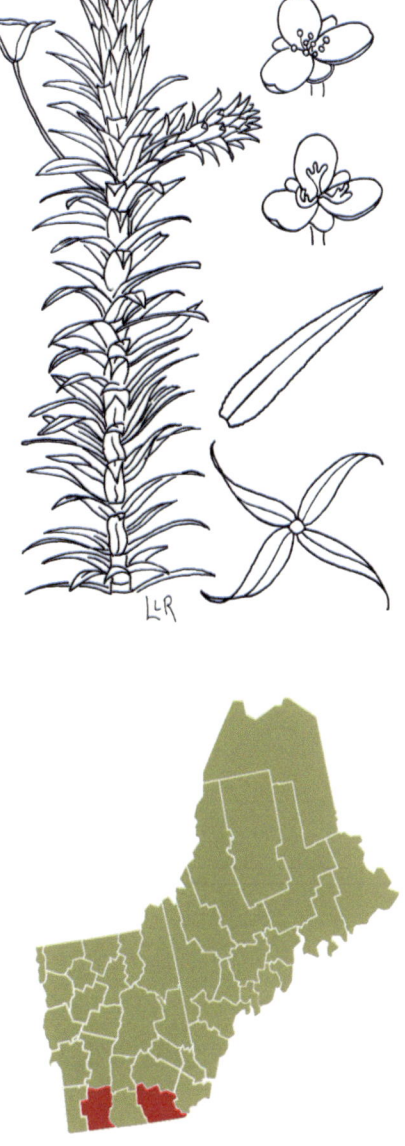

Eichhornia — Pontederiaceae
Water hyacinth

COMMON WATER HYACINTH   *Eichhornia crassipes* (Mart.) Solms.

Common Water hyacinth is a free-floating perennial with a buoyant cluster of spongy leaves above water surface and long, feathery, black roots hanging in water column. Leaves are in a basal rosette, green, glossy, and usually with greatly inflated stalks (3–33 cm long). Leaf blades are oval to round (2.5–11 cm long, 4–12 cm wide). Flowers are large (5–7 cm wide) and showy, blue or purplish-blue with a yellow spot, clustered in 4–15-flowered spikes, borne on erect stems (5–12.5 cm long) above the leaves.

HABITAT & DISTRIBUTION

Eichhornia crassipes is a non-native species and popular aquarium plant. It is uncommon in NH, found where plants escaped cultivation or were introduced during the same season. The species cannot tolerate winter in our area.

Notes: Native to South America, this is one of the most aggressive aquatic weeds known in warmer climates where it forms dense floating rafts. Its flowers open for one day only—within two hours after sunrise and then wilting by night.

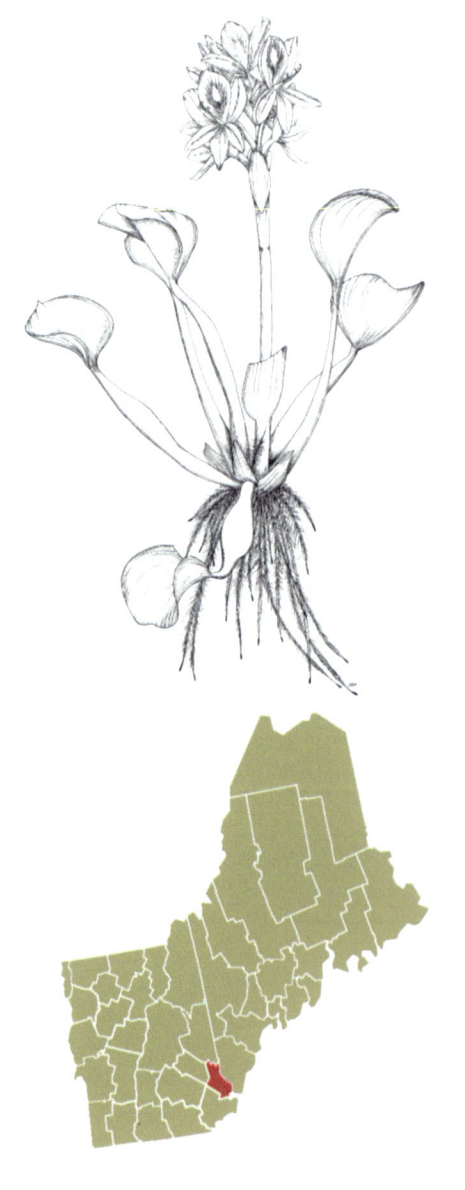

Elatine — Elatinaceae
Waterwort

The Waterworts are diminutive, mat-forming submersed plants, which can easily be overlooked. Individuals are sometimes emersed in mud when stranded by receding waters. Plants have simple, paired leaves along the stems and root at the lower nodes. Single, white or pinkish, bisexual flowers are borne in upper leaf axils which mature into capsules.

Three species are known in northern New England, including one non-native. All species are annuals and lack asexual reproduction, so population recruitment is strictly from seeds. Waterworts are popular aquarium plants.

KEY TO SPECIES

1. Leaves oblong and rounded, 1–5 mm long; flowers with 2 petals and stamens; seed surface with rounded pits; plants holding their shape when removed from water.......... *E. minima*
1. Leaves much longer than wide, 3–8 (–15) mm long; flowers with 3 petals and stamens; seed surface with longer 6-sided, angular-ended pits, plants limp and collapsing when out of water..2
   2. Leaf blades obovate or broadly spatulate, the larger blades 1.5–5 mm wide, tips rounded .... *E. americana*
   2. Leaf blades linear to egg-shaped, 0.5–3 mm wide, tips acute or sometimes notched....... .... *E. triandra*

E. minima

AMERICAN WATERWORT *Elatine americana* (Pursh) Arnott

American waterwort is an erect or prostrate herb with short (1.5–5 cm long), branched stems. Leaves are egg-shaped or broadly spoon-shaped, 3–8 mm long, larger blades 1.5–5 mm wide, with rounded tips. Minute flowers are borne singly in the leaf axils and develop into rounded capsules (1.5 mm wide). Seeds are cylindrical and marked on the surface with 6-sided, angular-ended pits.

HABITAT & DISTRIBUTION

Elatine americana is a native species in shallow (<1 m) waters and muddy shores of ponds, tidal or nontidal rivers. It is uncommon in northern New England. Rare and state-listed in both NH (*endangered*) and VT (*historical*).

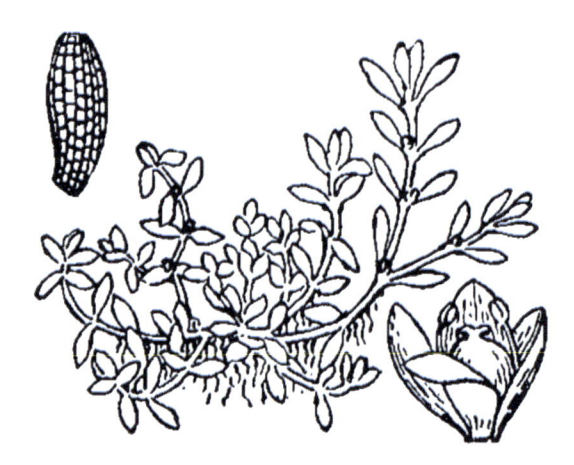

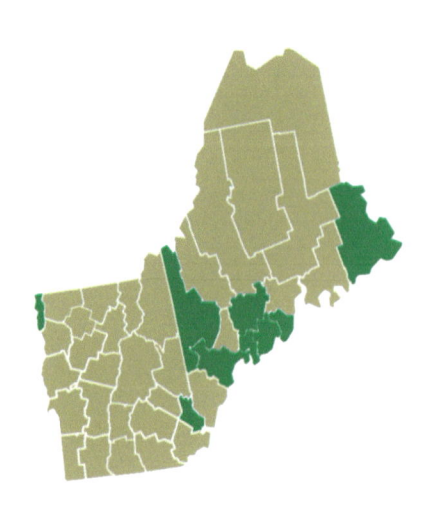

SMALL WATERWORT — *Elatine minima* (Nutt.) Fisch. & CA Mey.

Small waterwort is a small, erect, herb with short (to 3 cm long), sometimes few-branched, stems. Leaves are rounded and small (1–5 mm long). Minute flowers are borne singly in the leaf axils and develop into papery capsules containing 4–9 seeds. Seeds are cylindrical and marked on the surface with 8–9 rows of rounded pits.

HABITAT & DISTRIBUTION

Elatine minima grows in shallow (1–2 m) waters of lakes, ponds, tidal or nontidal streams, and exposed mudflats. It is native and widespread in ME and NH, but extremely rare in Vt. Flowers are self-pollinating, and seeds are dispersed by water. Colonies can become very dense (up to 2500 plants per m²) reaching highest densities in shallow water.

LONG-STEM WATERWORT — *Elatine triandra* Schkuhr

Long-stem waterwort is a prostrate herb with short (2–10 cm), highly branched, bright yellow-green stems. Leaves are linear to egg-shaped, 3–8 (–15) mm long, 0.5–3 mm wide, with acute or sometimes notched tips. Minute flowers are borne singly in the leaf axils and develop into papery capsules. Seeds are oblong and marked on the surface with 6-sided, angular-ended pits.

HABITAT & DISTRIBUTION

Elatine triandra is a non-native species in shallow waters (to 1 m) of lakes, ponds, tidal or nontidal streams, and mudflats. It is infrequent, or at least poorly documented, in northern New England. It is reported from the Connecticut River in MA. Introduced from Eurasia, where it is a weed in rice fields, this species is now widespread throughout North America. It can form dense carpets on lake bottoms and its seedbanks can be persistent.

Notes: It is very similar to Asian Waterwort (*E. ambigua* Wight), a non-native species, which has become introduced—most likely through aquarium disposal—into southern New England. Distinguishing between these two species is nearly impossible when plants are found submersed. When emersed, *E. ambigua* produces longer (0.5–2.5 mm), recurved fruit stalks (fruit stalks in *E. triandra* are ≤ 0.5 mm and erect). Recent molecular data confirm *E. ambigua* is present in waters of both CT and MA.

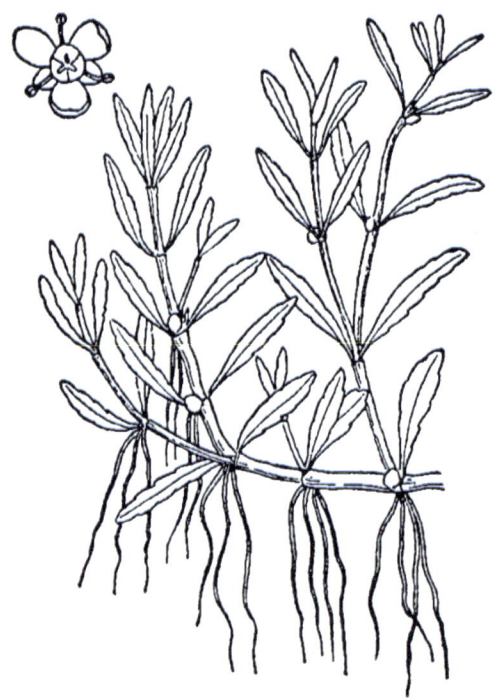

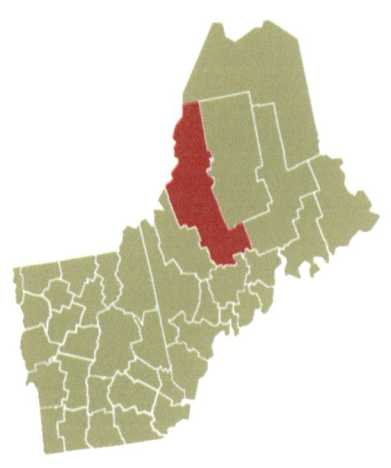

Eleocharis — Cyperaceae
Spikesedge

The Spikesedges, or spikerushes, are rooted, grass-like plants, with green upright stems and very reduced, inconspicuous leaves (represented by a closed, thin sheath at base of stems). Flowers are tiny, borne above water, in a single cluster (spikelet) of overlapping scales at the tip of a stem. Fruits are small and nut-like with bristles at the base, and a distinctive tubercle at top.

Many species of Spikesedge occupy shallow marshy wetlands and wet mud in northern New England. These species are more wholly emergent (and not detailed here). Three species, all native, can be found more submerged at times in ME, NH, or VT.

KEY TO SPECIES

1. Stem much narrower than the spikelet, stems generally round in cross-section, flexible, slender (<0.5 mm), typically short (5–8 cm tall), and often forming a dense turf.................2
1. Stem nearly as thick as the spikelet, flowering stems triangular in cross-section, stiff, 16–70 cm tall... *E. robbinsii*
    2. Plants of freshwater ponds and lakes (rarely brackish), without root-tubers at ends of rhizomes; nutlet with longitudinal rows of fine horizontal ridges .... *E. acicularis*
    2. Plants of brackish or saline coastal tidal marshes, with curved root-tubers at ends of rhizomes; nutlet without longitudinal rows of fine horizontal ridges .... *E. parvula*

E. acicularis

NEEDLE SPIKESEDGE *Eleocharis acicularis* (L.) Roem. & Schult.

Needle spikesedge is a delicate, rooted, perennial with white, threadlike rhizomes. Upright stems are clustered, flexible, short (5–8 cm tall), slender (<0.5 mm wide), and more slender than the spikelet. Spikelet is small (2–8 mm long) with dark brown scales with green mid-rib. Nutlet is 3-angled, with 2–4 bristles, and a tubercle separated from body by narrow constriction.

HABITAT & DISTRIBUTION

Eleocharis acicularis is a native species of shallow marshes and lake or river shores. It is common and widespread in northern New England. Plants can grow in dense clusters and form a turf. In deep water, plants may not produce flowers/fruits and can grow much taller (30+ cm).

Note: It closely resembles Little-headed spikesedge (*E. parvula*) of coastal saline or brackish tidal waters which are stouter and have curved, root-tubers on ends of the rhizome.

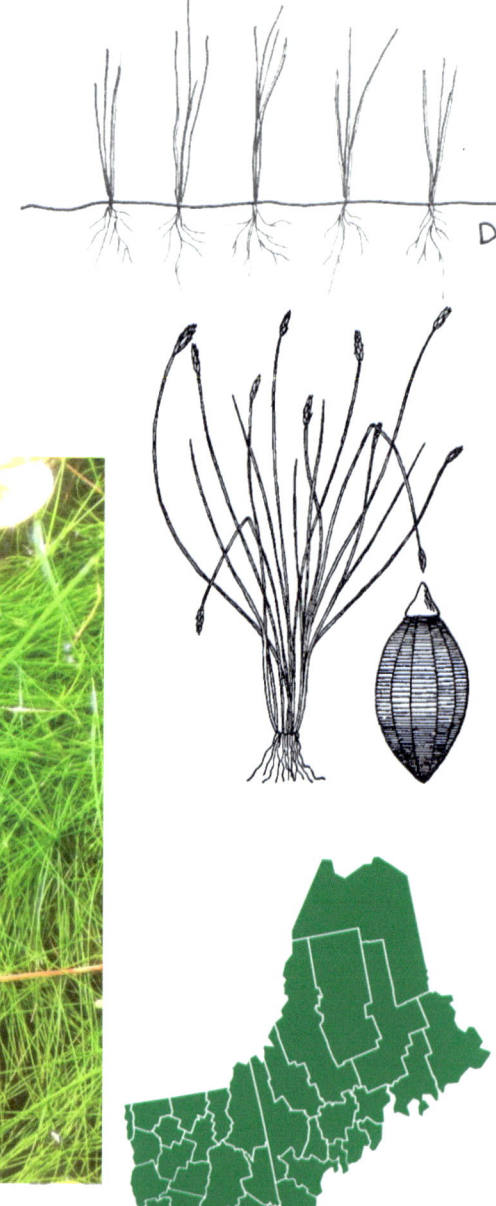

LITTLE-HEADED SPIKESEDGE *Eleocharis parvula* (Roem. & Schult.) Link

Little-headed spikesedge is a delicate, rooted, perennial with slender rhizomes terminated by pale, oblong, usually markedly curved, tubers (2–5 mm long). Upright stems are clustered, flexible, short (5–8 cm tall), and narrower than the spikelet. Spikelet is small (2–4 mm long) with light brown scales. Nutlet is 3-angled, lacking longitudinal rows of fine horizontal ridges, with 5–7 bristles, and a tubercle continuous with the fruit body.

HABITAT & DISTRIBUTION

Eleocharis parvula is a native species of coastal brackish or saline tidal waters in ME and NH. In deep water, plants may not produce flowers/fruits.

Note: It closely resembles Needle spikesedge (*E. acicularis*) of more freshwater habitats, which lacks root-tubers at ends of rhizomes.

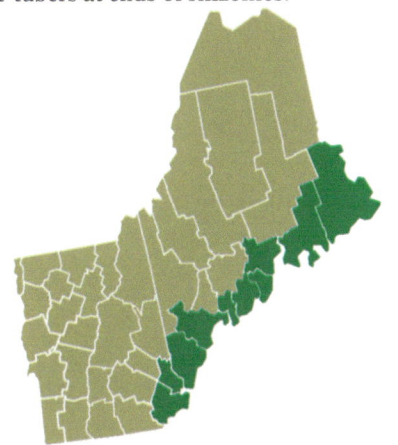

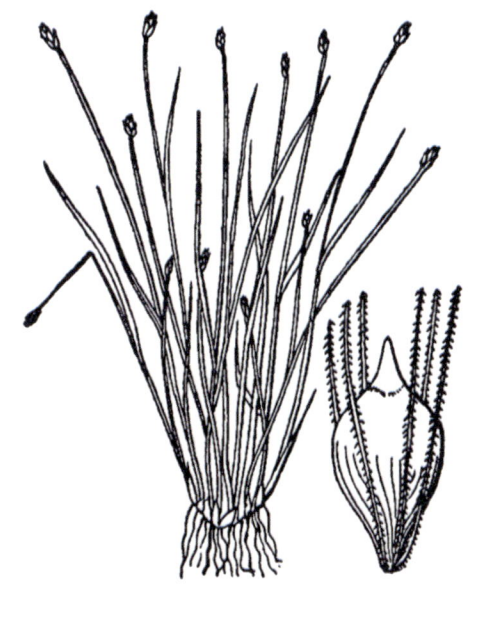

45

ROBBINS' SPIKESEDGE   *Eleocharis robbinsii* Oakes

Robbins' spikesedge is a rooted, perennial with buried, reddish, rhizomes (1–2 mm thick). Flowering stems are stiff, erect (16–70 cm tall, 1–2 mm wide), 3-sided, and nearly as thick as the spikelet. Numerous submersed, thread-like, flexible stems (resembling leaves) lacking spikelets are frequently present attached to base of plant. Spikelet is 9–33 mm long. Nutlet is 2-sided, with 6–7 bristles, and a distinct tubercle.

HABITAT & DISTRIBUTION

Eleocharis robbinsii is a native species of lakes and ponds where it forms clumps. It is widespread in ME and NH, but extremely rare in VT.

Note: It is similar to Water bulrush (*Schoenoplectus subterminalis*). Water bulrush is distinguished by alternate leaves and flowering stems with only their top portions held above water which have a single bract (2–20 mm long) attached below flower cluster appearing like a continuation of the stem.

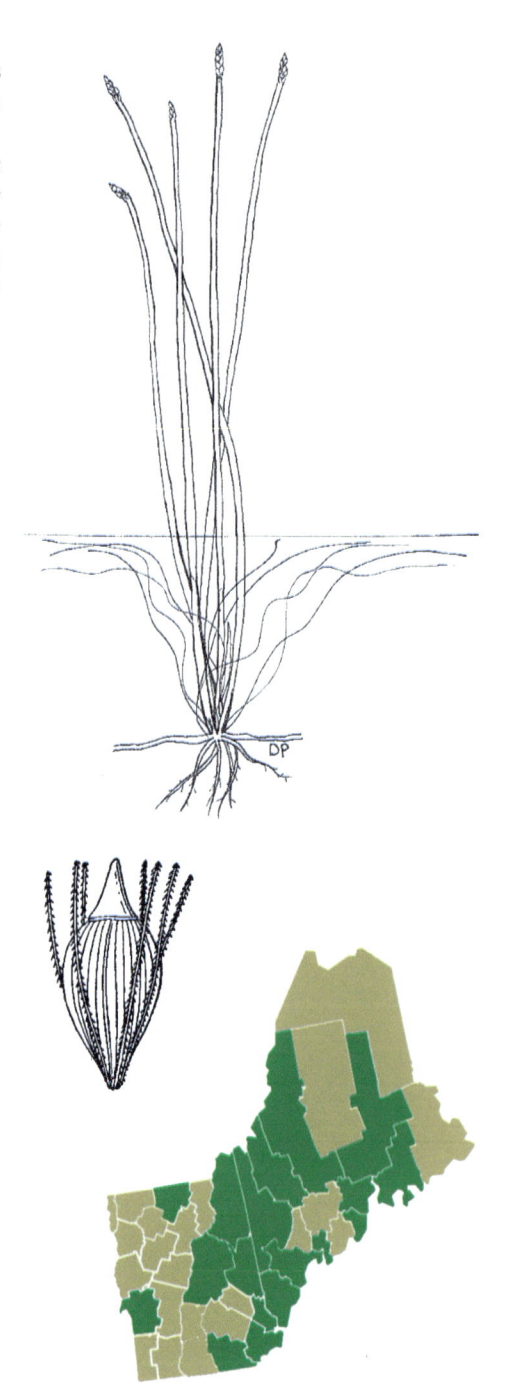

Elodea — Hydrocharitaceae
Waterweed

The Waterweeds are rooted, submersed aquatics with elongated, slender, branching stems which are somewhat brittle. Plants are leafy with small (<15 mm long, <5 mm wide), stalkless, bright green, recurved leaves borne in whorls of 2–3 per node. Flowers are unisexual (on different plants), white, and borne at the water surface.

Two species, both native and similar vegetatively, are known from northern New England. Waterweeds are commonly sold as aquarium "oxygenating" plants.

KEY TO SPECIES

1. Leaves oblong, mostly 2 mm wide (ranging 1–5 mm wide), firm, blunt-tipped, densely crowded and overlapping towards stem tips... *E. canadensis*
1. Leaves narrowly lance-shaped, mostly 1.3 mm wide (ranging 0.9–1.7 mm wide), weak, pointed, more or less equally spaced along stem.................................... *E. nuttallii*

E. canadensis

COMMON WATERWEED — *Elodea canadensis* Michx.

Common waterweed is a submersed, perennial with elongated (≥1 m), flexuous stems rooting from nodes. Leaves are mostly in whorls of 3, oblong, blunt tipped, mostly 2 mm wide (6–10 mm long, 1–5 mm wide), firm, and flat. Leaves in upper whorls can grow closely together and overlap. Flowers are unisexual (on separate plants), small (petals 2–5 mm long), borne singly, and white. Female flowers attached to long thread-like stalks at leaf axils, with sepals 2–3.5 mm long. Male plants are uncommon, but have flowers also elevated on long (2–30 cm) stalks.

HABITAT & DISTRIBUTION

Elodea canadensis is a native species of lakes, ponds, and slow rivers. It is widespread in northern New England but most common in neutral to calcareous waters. Plants grow rapidly in favorable conditions and form dense masses. It spreads by the detachment of resilient, densely leaved stem fragments (turions).

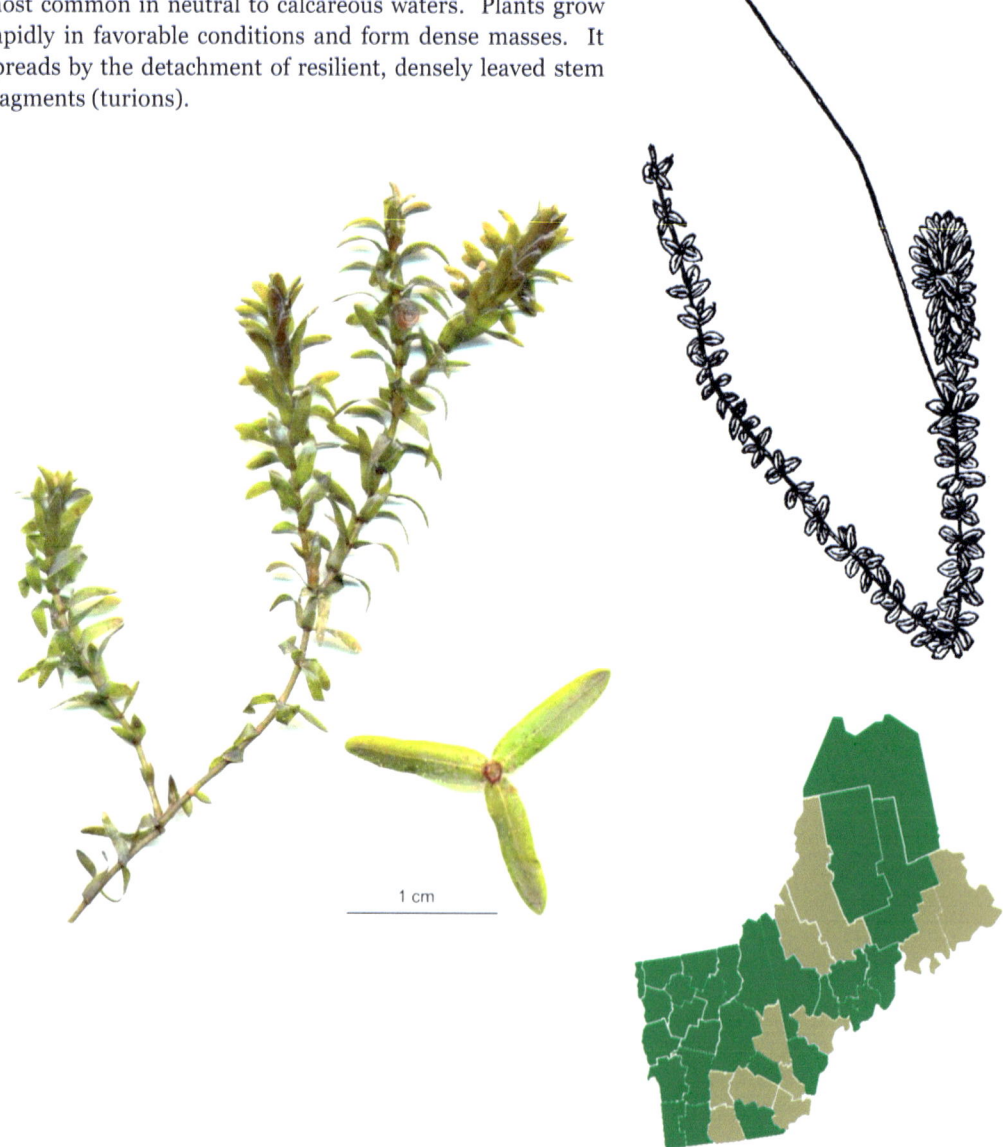

FREE-FLOWERED WATERWEED — *Elodea nuttallii* (Planch.) St. John

Free-flowered waterweed is a submersed, perennial, with elongated, flexuous stems rooting from nodes. Leaves are mostly in whorls of 3, narrowly lance-shaped, pointed, mostly 1.3 mm wide (4–15 mm long, 0.9–1.7 mm wide), weak, and edges slightly folded along midvein. Leaves more or less equally spaced along stem. Flowers are unisexual (on separate plants), small (petals 2.5 mm long), borne singly at leaf axils, and white. Female flowers attached to long thread-like stalks, with sepals 1–1.8 mm long. Male flowers are stalkless and released from plant to float on water surface.

HABITAT & DISTRIBUTION

Elodea nuttallii is a native species of lakes, ponds, slow rivers, and sometimes brackish tidal waters. It is widespread in northern New England, less common in VT. Plants are found in acidic to moderately alkaline waters.

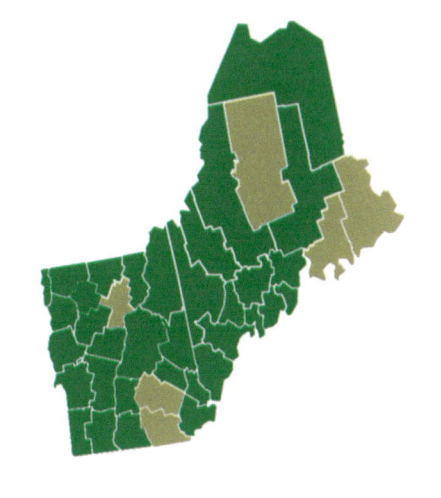

Eriocaulon — Eriocaulaceae
Pipewort

    The Pipeworts, or Hatpins, are diminutive, submerged, tufted, erect perennials with conspicuously white roots bearing distinctive cross-walls. Leaves are green, in dense, basal rosettes, narrow, tapered to a point from a flattened, pale, spongy base. Leaves are smooth, veiny, and somewhat translucent. Soft, flowering heads of minute, whitish flowers are borne at tips of erect, slender, longitudinally ridged stalks, and held above water surface.

    Two species occur in northern New England. Both are native and very similar in appearance.

KEY TO SPECIES

1. Flower cluster 4–6+ mm wide, white to gray-white, globose in shape, atop a 5–7 ridged stalk; usually a single flowering stalk per plant... *E. aquaticum*
1. Flower cluster 3–4 mm wide, dull gray or rarely yellow-brown, hemispherical in shape, atop a 4–5 ridged stalk; 1–4 flowering stalks per plant... *E. parkeri*

E. aquaticum

SEVEN-ANGLED PIPEWORT — *Eriocaulon aquaticum* (Hill) Druce

Seven-angled pipewort is a tufted, submersed plant with basal, green, tapered leaves, 1–10 cm long (to 40 cm), 2–5 mm wide. Flowering stalks (1 mm wide), usually one per leaf rosette, with 5–7 vertical ridges. Flower cluster 4–10 mm wide, white to gray-white, and globose in shape.

HABITAT & DISTRIBUTION

Eriocaulon aquaticum is a native plant of freshwater lakes, ponds, and slow rivers. It is widespread and common in northern New England. Plants are usually abundant on sandy substrates of shallow, low nutrient, acid waters.

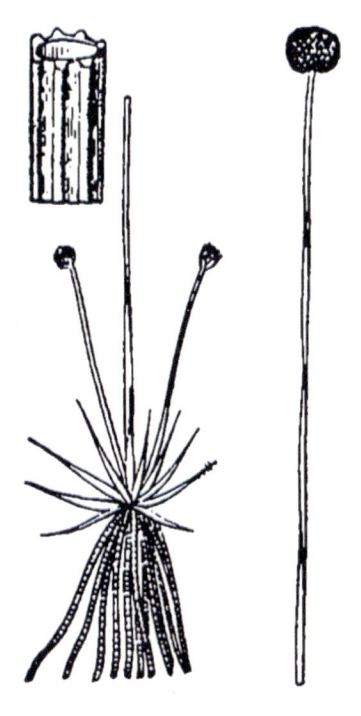

ESTUARY PIPEWORT *Eriocaulon parkeri* B.L. Robins.

Estuary pipewort, or Parker's pipewort, is a tufted submersed plant with basal, green, tapered leaves, 2–6 cm long, 1–2 mm wide. Flowering stalks, 1–4 per leaf rosette, are 2.5–10 cm long, with 4–5 vertical ridges. Flower cluster 3–4 mm wide, dull gray or rarely yellow-brown, and hemispherical in shape.

HABITAT & DISTRIBUTION

Eriocaulon parkeri is a native species of fresh to brackish tidal shores and open mudflats of estuaries. Restricted to coastal regions in ME where it is state-listed as *special concern*.

Notes: Plants are usually exposed at low tide, found at intertidal reaches to shores above high tide, and tolerant of oligohaline (salinity 0.5–5.0 ppt) waters.

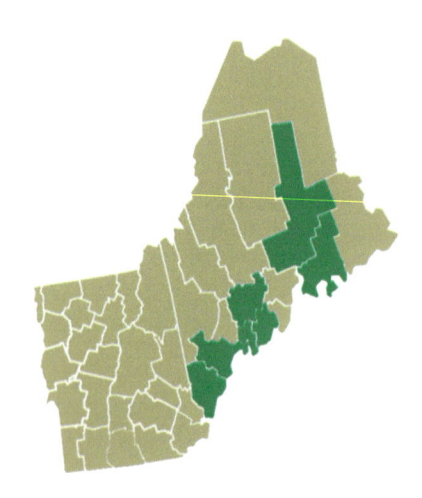

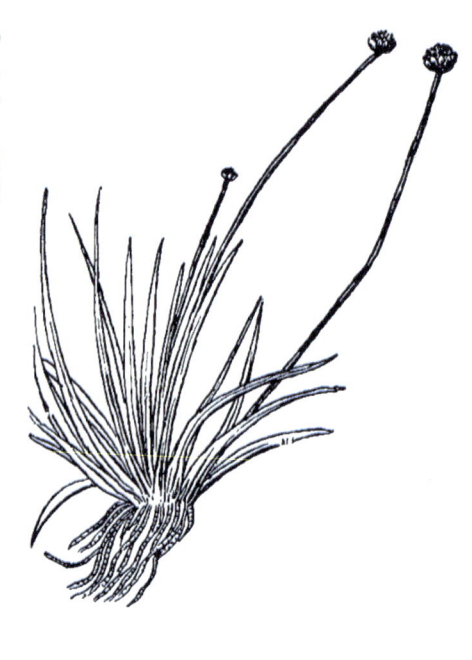

Glyceria — Poaceae
Manna grass

NORTHERN MANNA GRASS — *Glyceria borealis* (Nash) Batch.

Northern manna grass, or Small Floating manna grass, is a rooted, perennial grass (60–100 cm tall) with creeping rhizomes and narrow (1.5–5 mm), ascending, hairless stems. Leaves are typically floating on surface, especially early in season (becoming weakly erect with age). Leaf blades are linear (15–45 cm long, 2–7 mm wide), flat, abruptly tapered to a point, hairless, dark bluish-green, with a prominent midvein. Leaf blades are tubular at base and sheath the stem (sheaths, deep red when submerged, have fused margins for most of their length). Flower clusters are emergent, weakly erect (18–40 cm tall), several-branched, and borne from the uppermost leaf sheath. Flowers are enclosed in small (10–20 mm), linear, spikelets.

HABITAT & DISTRIBUTION

Glyceria borealis is a native grass of shallow lake and stream margins. It is widespread in northern New England. Mid-stem leaf blades have a dense covering of minute, pimply projections on their upper surface making them strongly hydrophobic.

Note: Can be confused with other uncommon *Glyceria* species which may have floating leaves, like Eastern manna grass (*G. septentrionalis* A.S. Hitchc.) and Sharp-scaled manna grass (*G. acutiflora* Torr.). These two plants (neither detailed here) differ mostly by finer details of flowers, are rare, and state-listed in Me, NH, and Vt.

54

Gratiola — Plantaginaceae
Hedge-hyssop

GOLDEN HEDGE-HYSSOP — *Gratiola aurea* Pursh.

Golden hedge-hyssop, or Golden pert, is a rooted, submersed, perennial with slender rhizomes. Plants are diminutive (1–5 cm tall), somewhat fleshy/stout, with smooth, somewhat four-angled, upright stems. Leaves are opposite, stalkless, small (≤5 mm long), narrow, awl-like and pointed, with toothless margins. Plants of this submersed life-form are sterile (vegetative only).

HABITAT & DISTRIBUTION
Gratiola aurea is a native species of low nutrient, mostly acidic, lake waters with full sun exposure. It is widespread in northern New England. Plants can be locally abundant, forming dense, turf-like, beds in either shallow or deep (4 m) waters.

Notes: The dwarfed, truly aquatic life-form (f. *pusilla* Fassett) is described here. At many occurrences, the species instead exists as a larger, fertile, emergent life-form on wet sandy shores of marshes, lakes and rivers. The emergent form has taller stems (10–30 cm) with gland-tipped hairs, larger leaves (1–3 cm long) with conspicuous glandular dots, and bright yellow flowers (12–16 mm long) borne singly in leaf axils.

The submersed form is similar in appearance to Waterwort (*Elatine*) which has rounded leaves and evident flowers/fruits.

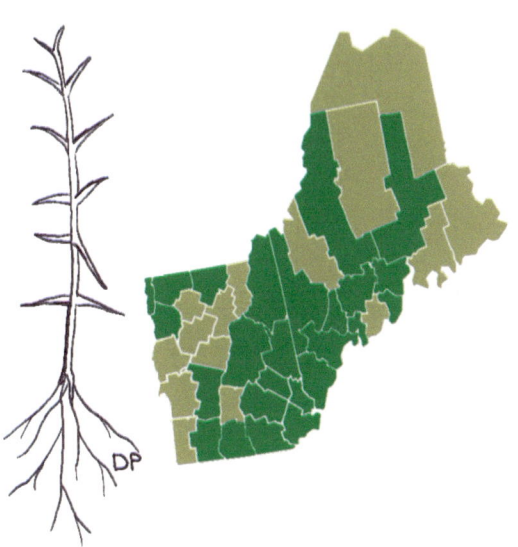

Heteranthera — Pontederiaceae
Mud-plantain

GRASS-LEAVED MUD-PLANTAIN   *Heteranthera dubia* (Jacq.) MacM.

Grass-leaved mud-plantain, or Water star-grass, is a submersed (or emersed on exposed mud), rooted, perennial with ribbon-like leaves, buried horizontal stems, and elongated, flexible, branching stems growing to water surface. Leaves are alternate, submersed, linear (3–10 cm long, 1–5 mm wide), thin, lacking an obvious mid-rib, and blunt-tipped. Flowers are yellow, with six linear (4–11 mm long) petal-like lobes, three yellow stamens, and borne singly at water surface. Each flower blooms for 1-day, opening within two hours of dawn, and wilting by dusk. Fruit an elongated (10 mm) capsule.

HABITAT & DISTRIBUTION

Heteranthera dubia is found in sluggish streams, shallow ponds, and lakes. It is native and most common in hard waters of western VT. It is rare and state-listed in ME (*special concern*) and NH (*threatened*).

Notes: Plants exhibit most extensive growth in alkaline waters and can become weedy. When not flowering, this species resembles narrow-leaved Pondweed (*Potamogeton*) species. It exhibits great variation in leaf shape and size, whereby plants in ponds tend to produce longer, narrower leaves than those in rivers.

Hippuris — Plantaginaceae
Mare's tail

COMMON MARE'S-TAIL *Hippuris vulgaris* L.

Common Mare's tail, or Bottlebrush, submersed rooted perennial with creeping rhizomes bearing stout, hollow, unbranched stems (25–100 cm long) which may be emergent at upper end. Leaves are whorled (6–12 per whorl), submersed, limp, ribbon-like (3–5 cm long, 1–3 mm wide), toothless, and pointed at tip. Upper stem segments that are emergent have fewer, shorter (0.5–3.5 cm), stiffer, and waxy leaves. Flowers are tiny (1 mm), green or purplish, and borne singly in leaf axils of emergent portions of stems.

HABITAT & DISTRIBUTION

Hippuris vulgaris is a native species of shallow, standing water of pools and ponds or slow streams. It is a northern, cold-water species at the southern end of its range in New England. It is rare and state-listed in ME (*special concern*), NH (*threatened*), and VT (*endangered*).

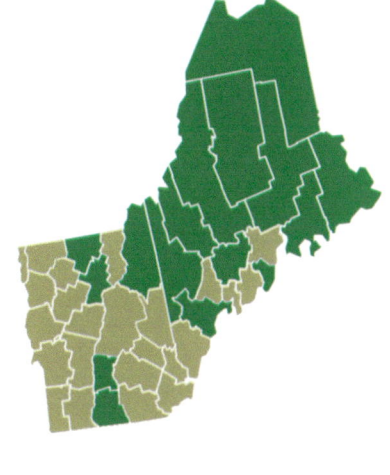

Hottonia — Primulaceae
Featherfoil

The Featherfoils are submerged (or sometimes stranded on shorelines), erect, annuals or perennials, with flattened, feathery leaves. Leaves are green, alternate, or sometimes clustered in whorl-like rosettes. Flowers are 5-lobed, white to violet, clustered in widely spaced whorls on erect stalks held above water surface. Fruit is a small (1.5–3 mm), globose, capsule.

Two species occur in northern New England, one introduced. Both species are exceptionally early-bloomers, bearing flowers from May to June.

KEY TO SPECIES

1. Flowers inconspicuous (2.5 mm wide), white, on short stalks (2–5 mm), clustered on a rosette of inflated, hollow flowering stems; plants usually absent during summer months.... .... *H. inflata*
1. Flowers showy (15–25 mm wide), white-violet, on long stalks (12–30 mm), clustered on a single, erect flowering stem; plants present during summer months................. *H. palustris*

H. palustris

AMERICAN FEATHERFOIL *Hottonia inflata* Ell.

American featherfoil is a submersed, short-lived annual with long (30–50 cm), flaccid, hollow stems, rooted at base. Leaves are alternate (or whorled/clustered near top), submersed (or floating at surface), oblong (2–8 cm long), feathery, and finely divided into linear, flattened segments. Flower clusters are floating on water surface, with swollen, air-filled, spongy stalks (to 30 cm long) bearing whorls of (3–10) inconspicuous white flowers.

HABITAT & DISTRIBUTION

Hottonia inflata is a native Coastal Plain species found in still, acidic waters of ponds, swamps, and ditches. It is rare and state-listed in ME (*threatened*) and NH (*endangered*).

Note: This odd-looking plant is unmistakable in spring with its buoyant, swollen, flower clusters. These clusters can detach and become free-floating. As a "winter annual" species, after fruiting plants typically wither and die by mid-July. Abundance at a site can vary greatly from year to year as a function of precipitation.

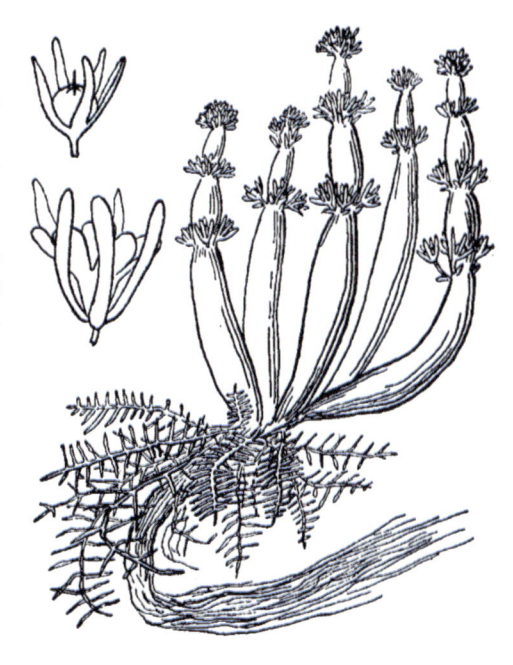

WATER VIOLET *Hottonia palustris* L.

Water violet, or European featherfoil, is a submersed, perennial with long (to 90 cm), flaccid, stems, rooted at nodes and base. Leaves are alternate (or whorled/clustered near top), submersed (or floating at surface), oblong (2–13 cm long), feathery, and finely divided into linear, flattened segments. Flower clusters are held above water surface, on long (to 40 cm), erect, finely haired, stems bearing whorls of 3–7 conspicuous flowers (15–25 mm wide). Flowers white to violet, with yellow centers, on stalks 12–30 mm long.

HABITAT & DISTRIBUTION

Hottonia palustris is a native of Europe and north Asia, found in still, acidic waters of ponds. It was probably **introduced** through water garden cultivation. It is uncommon in lakes region of NH but could become weedy and problematic.

Notes: Unlike our native American featherfoil, this species is perennial (and should be present thoughout growing season), has showy flowers, and lacks the rosettes of inflated flowering stems. When not in flower, its leaves can be mistaken for *Proserpinaca* or *Myriophyllum*.

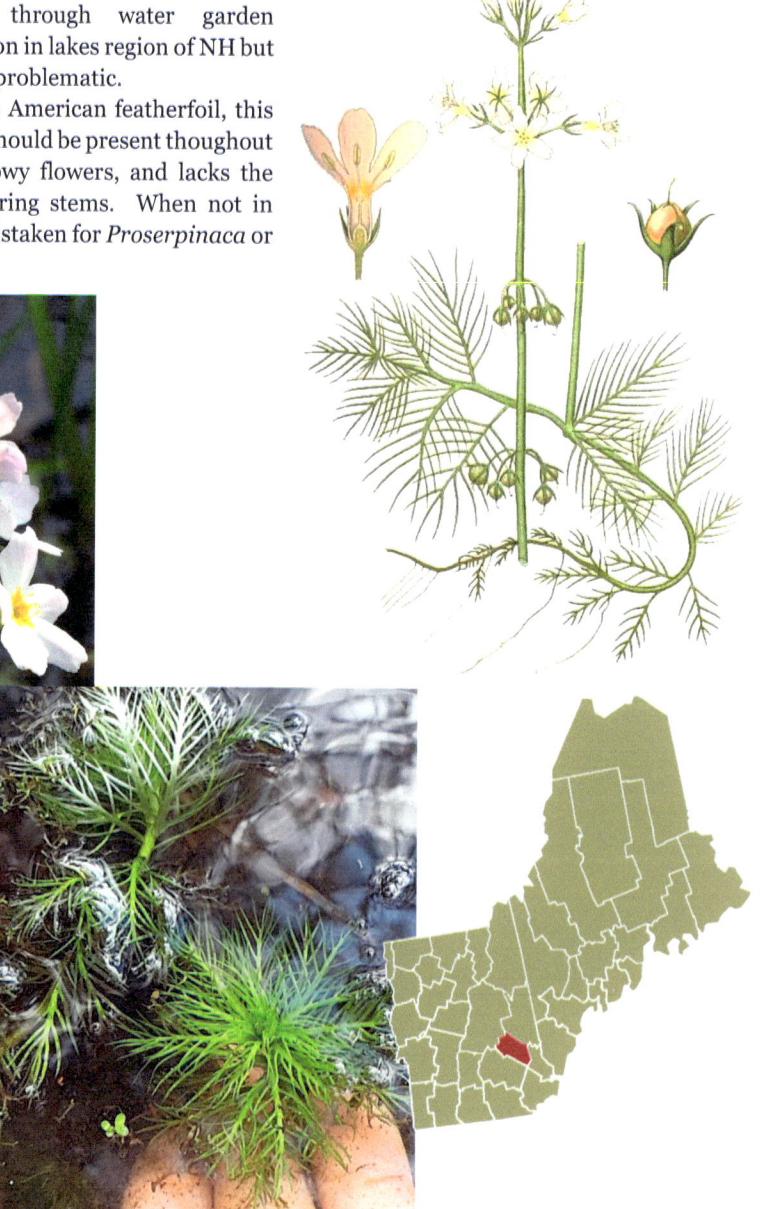

60

Hydrilla — Hydrocharitaceae
Hydrilla

HYDRILLA — *Hydrilla verticillata* (L.f.) Royle

Hydrilla, or Water-thyme, is a submersed, perennial with rhizomes and elongated, flexuous, branched stems. Leaves are in whorls of 4–8, linear (8–15 mm long, 1–4 mm wide), sharply pointed, with saw-toothed margins, and bottom surface with minute prickles or bumps along midvein. Flowers are unisexual (on same or separate plants), small, white, and borne singly from upper leaf axils. Female flowers with elongated (10–50 mm) floral tubes reaching water surface. Fruit a small (4–6 mm long) capsule.

HABITAT & DISTRIBUTION
Hydrilla verticillata is a non-native species of lakes, ponds, and rivers. It is known from eastern ME where it is **invasive** and prohibited. Native to Asia, it was introduced into the US as an aquarium plant. It produces two types of overwintering turions: underground, pale brown, tuber-like ones and dark green, stiff-scaled ones on upright stems.

Notes: A problematic weed that is fast-growing, easily and rapidly dispersed, and with remarkably broad ecological tolerances (of water depth, pH, salinity, and nutrient levels). Anywhere it is found it poses a serious threat to the waterbody. It may be confused with Waterweeds (*Elodea*).

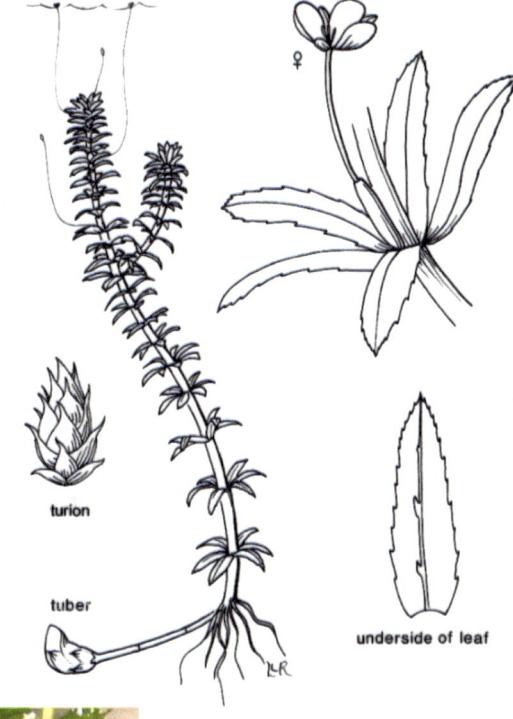

Hydrocharis — Hydrocharitaceae
Frog-bit

EUROPEAN FROG-BIT *Hydrocharis morsus-ranae* L.

European frog-bit is a perennial, free-floating or rooted floating-leaved plant, with horizontal stoloniferous stems and branched roots. Leaves are in basal rosettes and floating with long stalks (4–6 cm long) and a slight spongy central region on lower blade surface. Leaf blades are leathery, broad (1.5–6.5 cm long, 1–6 cm wide), heart shaped to near circular (with 5–7 broadly arching veins), with rounded tips. Each leaf with two stipules at base of leaf stalk. Flowers are unisexual (on different plants) with three whitish petals (yellowish at base) and held above water. Fruits, if produced, round (4–12 mm wide), berry-like, on recurved stalks. Turions develop at end of stolons in late summer.

HABITAT & DISTRIBUTION

Hydrocharis morsus-ranae is a non-native species of quiet, shallow lakes, sheltered pools, and slow river margins. It is known from ME and western VT where it is **invasive** and prohibited. Native to Eurasia, this species was introduced into North America as a water garden escape.

Notes: A problematic weed that can develop dense, floating mats easily dispersed by fragmentation. Individual plants are known to form up to 100–150 turions, which can remain dormant for 2+ years.

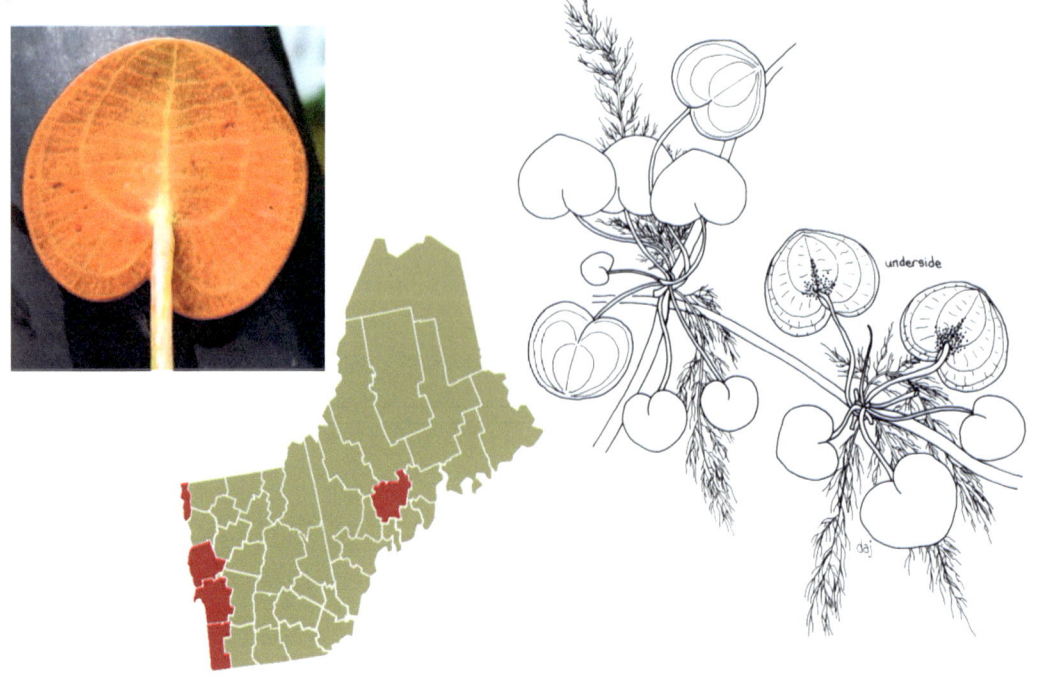

Hypericum — Hypericaceae
St. John's-wort

NORTHERN ST. JOHN'S-WORT — *Hypericum boreale* (Britt.) Bickn.

Northern St. John's-wort is a delicate, submersed, rooted, perennial, with slender rhizomes and flaccid, elongated, mostly unbranched stems (1–2 mm wide). Leaves are opposite, small (6 mm long), rounded, toothless, 3-veined, and without stalks (round bases of blade clasping stems). Plants usually lack flowers.

HABITAT & DISTRIBUTION
Hypericum boreale is a native species that occurs in shallow ponds, streams, and marshes. It is widespread in northern New England.

Notes: The flexuous, sterile, submersed life-form (f. *callitrichoides* Fassett) is described here. In many cases, the species also occurs nearby as the typical, fertile, emergent life-form on wet shores or marshes. Emergent plants have erect stems (10–40 cm tall), larger, more oblong, leaf blades (1–3 cm long) with scattered translucent dots on surface and 3–5 veins, and small, yellow flowers borne near tips of spreading upper branches. The submersed life-form is similar to Water starworts (*Callitriche*).

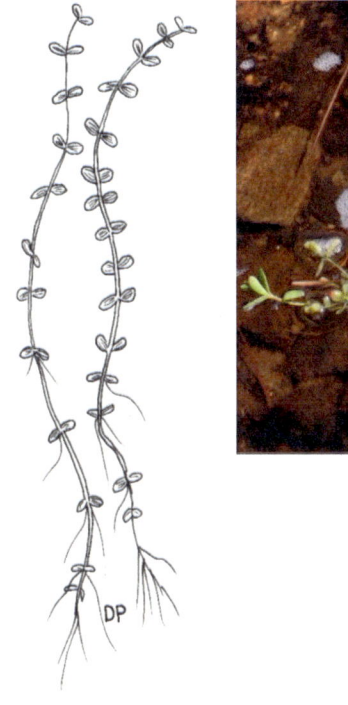

Isoëtes — Isoëtaceae
Quillwort

The Quillworts are tufted, submersed, perennial, spore-producing (i.e., non-flowering) plants with grass-like leaves arranged in rosettes, and rooted below a bi-lobed stem. Leaves are linear, green, with four hollow lengthwise chambers, scattered cross-walls, and swollen, whitish bases where the spore-producing structures (sporangia) are enclosed.

Vegetatively, most species are very similar in appearance. Species identification is challenging and requires microscopic examination of sculptural details of mature macrospores. These diagnostic, larger, spherical spores (bone-white when dry), usually borne in bases of outer leaves, are typically available from our plants only after July. A thin, membranous flap of tissue (termed a velum) covers the inside surface of the sporangium. The degree of velum coverage (e.g. completely or partially) is also informative and can be determined by lifting it off the sporangium wall with forceps.

Nine species occur in northern New England, all native. Many species are known to hybridize frequently.

KEY TO SPECIES

1. Velum covering the entire sporangium; sporangium wall unpigmented/translucent; leaves straight and brittle..*I. prototypus*
1. Velum covering 50% or less of sporangium; sporangium wall usually brown pigmented (entirely, streaked, or spotted); leaves straight or curved, flexible.............................2
    2. Velum covering about half of sporangium; leaves short (≤5 cm)............. *I. viridimontana*
    2. Velum covering less than half of sporangium; leaves longer (5–50 cm)...................3
3. Megaspores densely covered with thin, sharp, spine-like projections; leaves deciduous....... .... *I. echinospora*
3. Megaspores covered with ridges; leaves evergreen, lasting 1+ years..4
    4. Megaspore diameter averaging <0.5 mm, patterned with an unbroken network of ridges.. .... *I. engelmannii*
    4. Megaspore diameter averaging >0.5 mm, patterned with crest-like, wrinkled or network of broken, unconnected ridges.. 5
5. Leaves light green with abundant stomates; plants occasionally emergent; megaspores girdle (area just below equatorial ridge) obscure... *I. septentrionalis*
5. Leaves dark green to red-brown with few or no stomates; plants always submerged; megaspore girdle evident.. 6
    6. Megaspore diameter averaging >0.6 mm; leaves coarse and firm, abruptly tapering to the tip; plants typically in deep (1–3+ m) water...7
    6. Megaspore diameter averaging <0.6 mm; leaves soft and flaccid, gradually tapering to the tip; plants typically in shallower (0.5–1 m) water... 8
7. Megaspores with sharp or roughened ridges, girdle densely covered with small bumpy projections.. *I. lacustris*
7. Megaspores bearing smooth, rounded ridges and a smooth girdle............... *I. hieroglyphica*
    8. Megaspores with girdle densely covered with small bumpy projections, ridges having irregular and roughened crests.. *I. tuckermanii*
    8. Megaspores with a smooth girdle, ridges having rounded and smooth crests...*I. acadiensis*

I. lacustris

I. echinospora

ACADIAN QUILLWORT *Isoëtes acadiensis* Kott

    Leaves are 9–35 per plant, dark green to red-brown with few or no stomates, gradually tapering to the tip, mostly 5–20 cm long, 0.25–1.5 mm wide, and mostly recurved. Plants always submerged, usually in shallow (0.5–2 m deep) water. It is very rare and state-listed as *special concern* in ME and *endangered* (historical) in NH. Restricted to acidic, low nutrient ponds and lakes or sluggish streams.

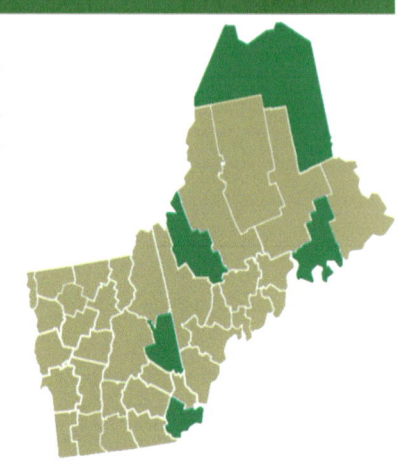

SPINY-SPORED QUILLWORT *Isoëtes echinospora* Durieu

    Leaves are mostly 5–15 cm (to 40 cm) long, 0.5–1.5 mm wide, with slender tips. Leaves usually stiff in ponds and soft in flowing water. The most common and widespread species in northern New England occupying clear, low nutrient, acidic lakes, ponds, or streams. Our plants are ssp. *muricata* (Durieu) A. & D. Löve.

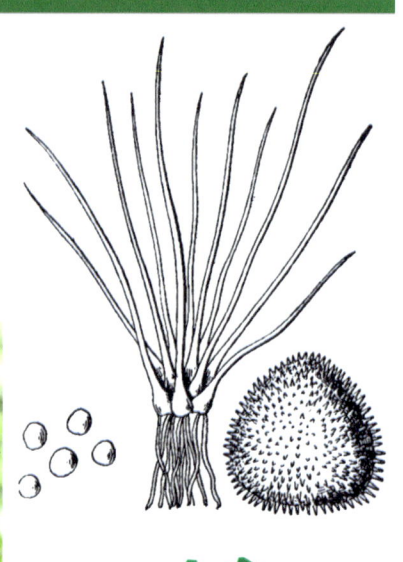

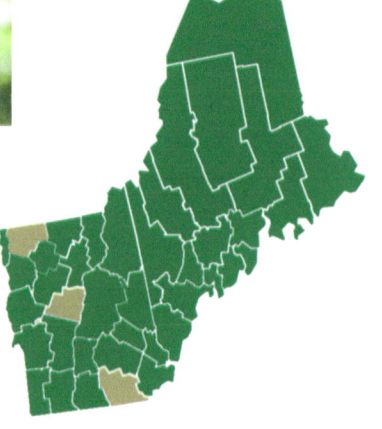

ENGELMANN'S QUILLWORT — *Isoëtes engelmannii* A. Braun

Leaves are 6–50 cm long, 0.5–2 mm wide. Usually in shallow ponds and slow streams with fluctuating waters. Plants can be emersed later in season as. It is restricted to southern areas in our region where it is rare and state-listed as *endangered* in NH and *threatened* in VT.

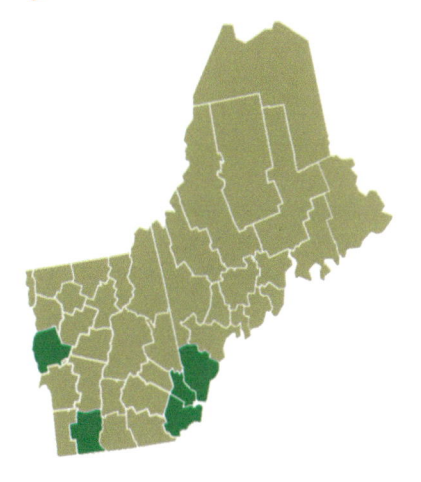

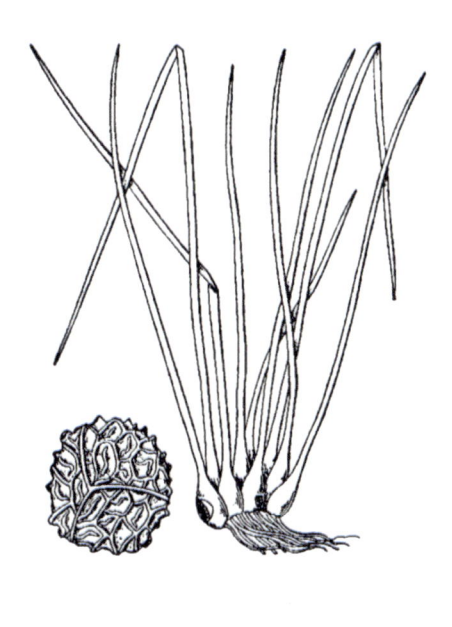

CARVED QUILLWORT — *Isoëtes hieroglyphica* A.A. Eaton

Leaves are evergreen, dark green, coarse and firm, short (5–10 cm long, 1–2 mm wide), abruptly tapering to the tip. Plants are submerged in deep, cold, clear, gravelly lakes, ponds and streams. Restricted to northern ME.

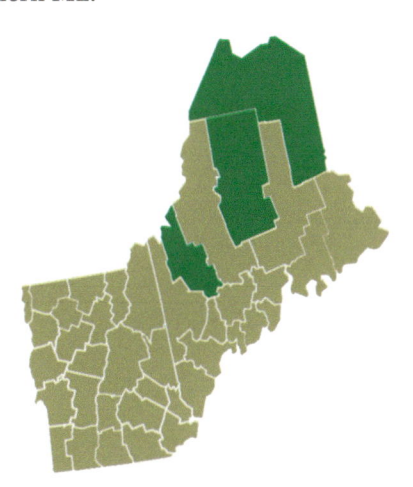

LAKE QUILLWORT — *Isoëtes lacustris* L.

Leaves are dark green, coarse and firm, short (5–10 cm long, 1–2 mm wide), abruptly tapering to the tip, with few or no stomates. Plants are submerged in deep (1–3+ m), cold, clear lakes, ponds and streams. It is very rare and state-listed as *endangered* in NH and *rare* in VT.

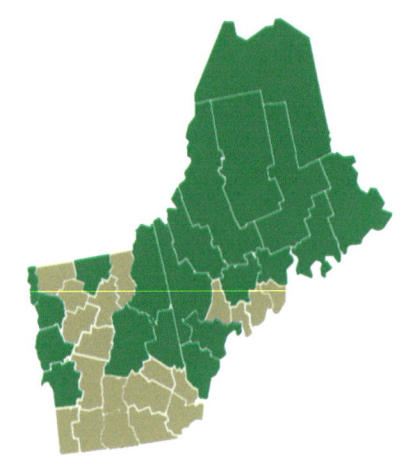

PROTOTYPE QUILLWORT — *Isoëtes prototypus* D.M. Britt.

Leaves 10–25+ very straight (5–12 cm long), rigid and brittle, short, gradually tapering to the tip. Leaves are dark green, except for reddish brown- or chestnut-colored bases, often breaking (i.e., not flexing) when applying pressure downward on the tips of the leaves. Plants are submerged in deep, cold, clear, acidic lakes with sandy substrates. Restricted to ME where it is rare and state-listed as *threatened*.

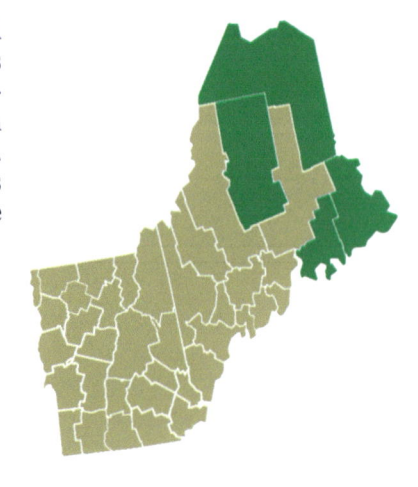

NORTHERN SHORE QUILLWORT   *Isoëtes septentrionalis* D.F. Brunt.

Leaves are bright green to yellow-green (whitish-green at base), 10–20 cm long, 0.5–1.5 mm wide, with abundant stomates. Plants usually submersed, occasionally emergent, in shallow waters (<0.5m deep) of streams and lakes with a sterile sand, gravel, boulder substrate. It is rare in northern New England where it is state-listed in ME (*endangered*) and NH (*endangered*).

Note: Our northern plants were formerly recognized as a variant ("var. *canadensis*") of *I. riparia* Engelm. ex A. Braun.

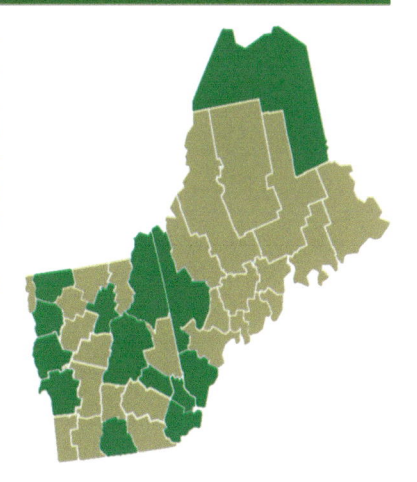

TUCKERMAN'S QUILLWORT   *Isoëtes tuckermanii* A. Braun ex Engelm.

Leaves are thin, soft and flaccid, mostly 5–20+ cm long, 0.25–1.5 mm wide, gradually tapering to the tip. Plants always submerged, in ponds and lakes 1 (–3) m deep. It is widespread in northern New England, but rare in VT.

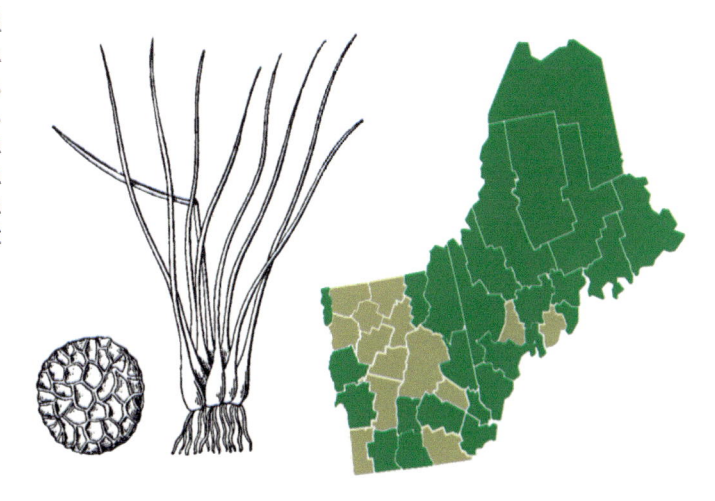

GREEN MOUNTAIN QUILLWORT   *Isoëtes viridimontana* Rosenthal & Taylor

Plants submerged, few-leaved (5-17), and leaves short (up to <5 cm long). This recently described species is known only from a single acidic pond in VT, where it is state-listed as *endangered*.

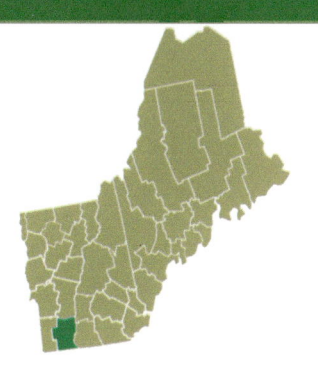

Juncus — Juncaceae
Rush

BAYONET RUSH — *Juncus militaris* Bigelow

Bayonet rush is a rooted, grass-like perennial, with submersed rhizomes and stout, erect stems. Stems are tall (to 150 cm) and round or oval in cross-section (5–12 mm wide). Leaves are of two types: many, submerged (or tips floating), flaccid, thread-like leaves attached to rhizomes, and usually a single, emersed, stiff leaf attached to the erect stem. The emersed leaf is stout (50–100 cm long, 2–4 mm wide), round or oval in cross-section, with hardened internal cross-walls, pointed, attached near middle of stem, and overtopping the flower cluster. Flowers are minute, green or brown, borne in a multi-branched cluster (4–15 cm tall) at stem tip. Fruit is a small (2–3 mm long), 3-seeded capsule.

HABITAT & DISTRIBUTION
Juncus militaris is a native species of shallow lakes and rivers. It is scattered mostly in eastern regions, rare and state-listed as *endangered* in VT. The submersed leaves are found particularly in deeper waters, where plants may remain sterile. Plants can form dense colonies.

Notes: The hardened partitions in leaves, present at regular intervals, can be felt by pulling the leaf between your fingers and pressing gently. Similar to Robbins' spikerush (*Eleocharis robbinsii*) and Water bulrush (*Schoenoplectus subterminalis*), but neither possess emersed leaves or internal cross-walls.

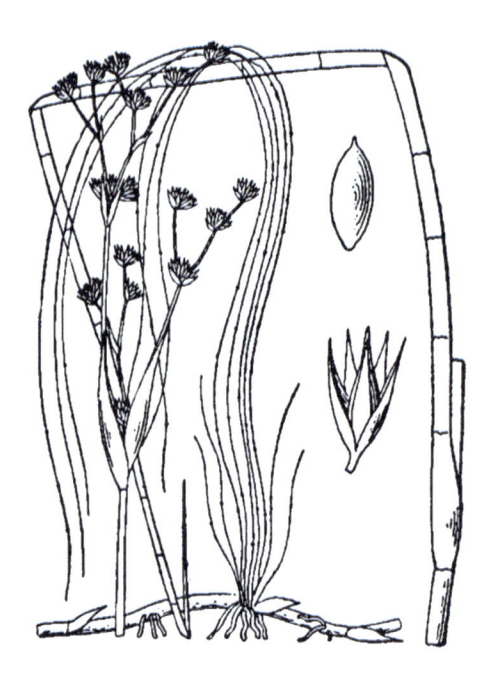

Lemna — Araceae
Duckweed

The Duckweeds are small (<15 mm long), green, perennials, mostly floating on surface of quiet waters. Plant body (or "thallus") is greatly reduced (not differentiated into stem and leaf), flattened, oblong to narrowly egg-shaped, very obscurely 1–5 veined above, and one root below extending into water. Usually 1–several plants are connected. Flowers and fruits are tiny, rarely seen.

Three species occur in northern New England. Identification is best when plants are fresh. Individuals consistently reproduce vegetatively and are usually abundant, often intermixed with other duckweeds (*Spirodela* and *Wolffia*).

KEY TO SPECIES

1. Plant bodies 6–15 mm long, oblong to broadly lance-shaped in outline, tapering to a conspicuous narrow stalk by which mature plant bodies remain attached laterally to parents, floating just below water surface in tangled masses; root often absent.... *L. trisulca*
1. Plant bodies 1–6 mm long, round, egg-shaped, or narrowly oblong in outline, without stalks, usually floating on surface of water; root present..2
    2. Mature plants <0.5–1.5 mm wide, sides somewhat parallel, obscurely 1-veined (may appear unveined)... *L. valdiviana*
    2. Mature plants 1.3–3 mm wide, sides curved, with 3–5 obscure veins............... *L. minor*

L. minor

COMMON DUCKWEED — *Lemna minor* L.

Common duckweed is a diminutive, perennial, floating on water surface. Its body, or thallus is small (1–6 mm long, 1–3 mm wide), flattened, egg-shaped, with 3–5 very obscure veins on top surface, with or without a row of raised projections along midvein. Root is mostly over 3.5 cm long (to 15 cm), with a rounded tip, extending into water. Turions not produced.

HABITAT & DISTRIBUTION

Lemna minor is a native species of ponds, lakes, marshes and streams. It is common and widespread in northern New England.

Notes: Similar to Turion duckweed (*L. turionifera* Landolt) which is now considered *historic* in VT, once known from Chittenden & Windsor Cos. Turion duckweed differs by having an intense reddish color on lower surface, a distinct row of minute projections along midvein, and sometimes small (0.8–1.6 mm wide), brownish, rootless, turions. Also similar to the rare Minute duckweed (*L. perpusilla* Torr.), once known from Chittenden Co., VT but now considered *historic*. It differs by having much shorter (<3.5 cm long), pointed roots, a body with a prominent, raised projection at tip, and a winged root sheath.

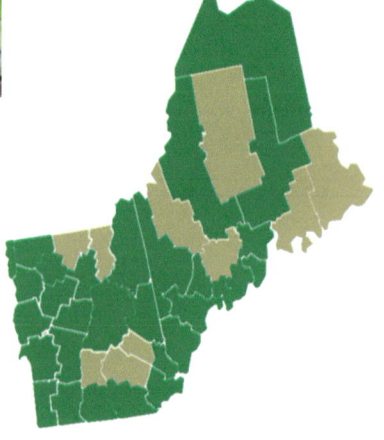

IVY-LEAVED DUCKWEED — *Lemna trisulca* L.

Ivy-leaved duckweed is a diminutive perennial, suspended just below water surface. Its body, or thallus, is 6–15 mm long (1.4–4 mm wide), flattened, oblong to broadly lanceolate in outline, with a somewhat toothed margin towards tip, tapering at base to a conspicuous, narrow stalk. Root deciduous and sometime absent. Mature plant bodies remain attached to parents via stalks.

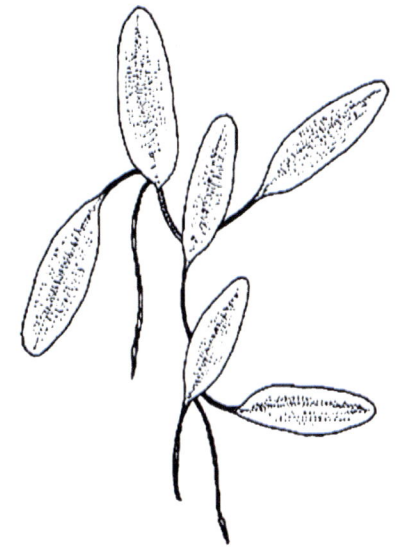

HABITAT & DISTRIBUTION

Lemna trisulca is a native species of ponds and lakes, most common in alkaline waters. It is absent in most of northern New England. Rare and state-listed as *endangered* in NH.

Notes: One of the more distinctive duckweeds with its long stalks. This species often forms extensive, tangled colonies where 3–30 clones can be interconnected. Bodies become shorter and thicker when flowering.

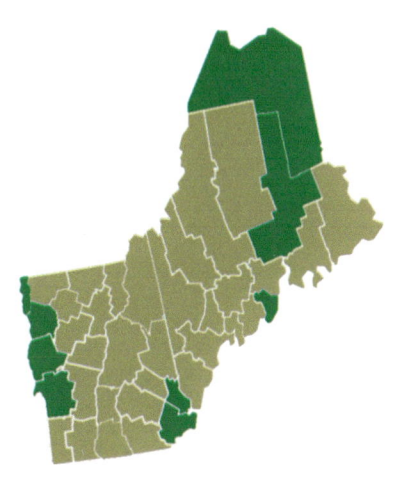

PALE DUCKWEED   *Lemna valdiviana* Phil.

Pale duckweed is a diminutive, perennial, floating on water surface. Its body, or thallus is small (2.5–5 mm long, <0.5–1.5 mm wide), oblong, sides somewhat parallel, with 1 vein on top surface, with or without a row of raised projections along midvein. Root to 1.5 cm long, with a rounded to pointed tip, extending into water. Turions not produced.

HABITAT & DISTRIBUTION

Lemna valdiviana is a native species of ponds, lakes, marshes and streams. In northern New England is known only from NH where it is rare and state-listed as *endangered* (considered historical).

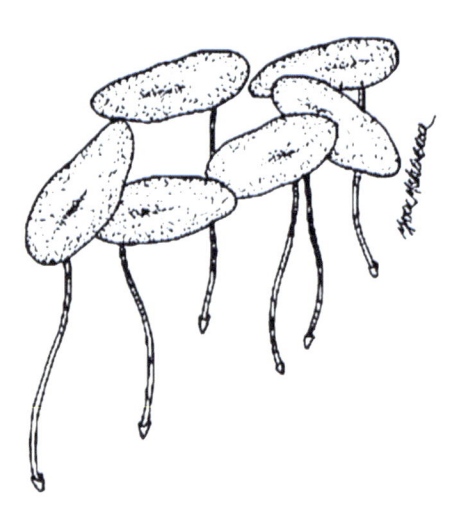

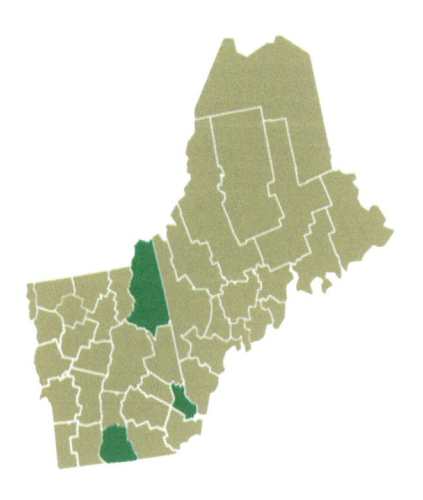

Lilaeopsis — Apiaceae
Grasswort

EASTERN GRASSWORT — *Lilaeopsis chinensis* (L.) Kuntze

Eastern grasswort is a low, submersed or emergent, perennial with creeping, horizontal rhizomes which root at nodes. Leaves are 1–few, arising from the rhizome, linear or spatulate in shape, without an expanded blade, 1–5 cm long, hollow, rounded to flattened, transversely divided with 4–6 septa. Flowers inconspicuous, bisexual, and borne 4–10 in clusters from base of leaves. Flower cluster stalks are about as long as or longer than the leaves. Fruits are small (<3mm), rounded nutlets with 5 corky ribs.

HABITAT & DISTRIBUTION
Lilaeopsis chinensis is a native species in shallow waters and exposed mudflats of freshwater to brackish marshes, tidal streams, canals, and ditches or saltmarshes. It is rare along southeastern coast, where it is state-listed in NH (*endangered*).

Notes: This is a coastal species typically occupying intertidal areas which become inundated at high tide. It can form dense colonies in sunny areas. Submersed plants may remain sterile.

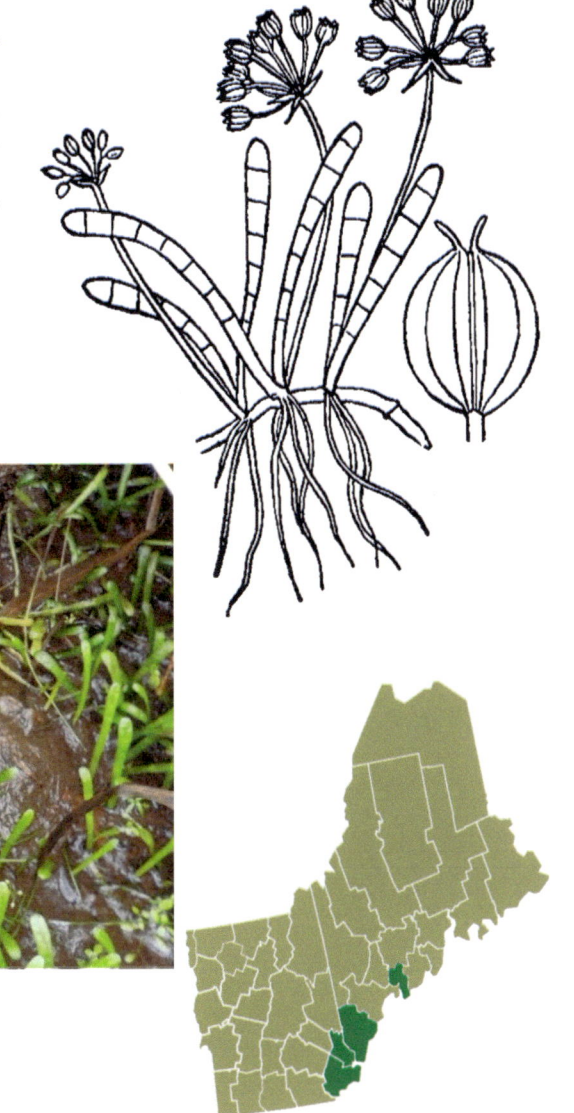

Limosella — Scrophulariaceae
Mudwort

ATLANTIC MUDWORT — *Limosella australis* R. Br.

Atlantic mudwort is a diminutive, rooted, submersed, herb with horizontal stoloniferous stems rooting at nodes. Leaves are fleshy, linear (2–5 cm long, 1–2 mm wide), round in cross-section, and tufted, 5–10 leaves clustered at bottom of plant. Flowers are small (3 mm wide), with 5 white petals (3.5–4 mm long) and 4 stamens, borne singly on erect stalks (about half of leaf length) that arise from nodes, several per plant. Fruit is a round capsule.

HABITAT & DISTRIBUTION
Limosella australis is a native species of shallow, tidally exposed, sand or mud of estuaries, rivers, and ponds. It is rare along the coast, where it is state-listed in both ME (*special concern*) and NH (*endangered*).

Notes: This succulent, sometimes mat-forming, plant is tolerant of brackish water (to 18 ppt salinity) and exposure at low tides. It is most common and abundant in shallower waters. Technically a perennial species that acts as an annual due to difficulty in surviving winter conditions.

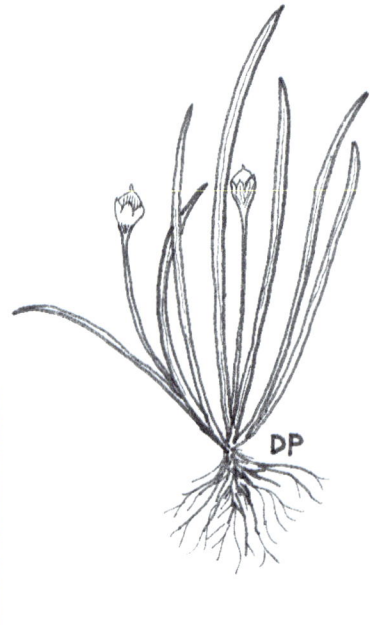

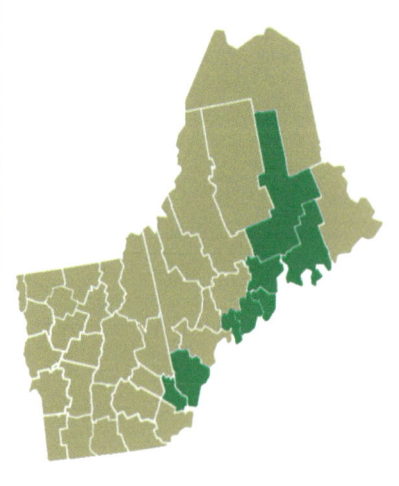

Littorella — Plantaginaceae
Shoregrass

AMERICAN SHOREGRASS — *Littorella americana* Fern.

American shoregrass is a small, submerged (sometimes emergent), rooted, perennial with a rosette of stiff, bright green leaves and delicate, white to green, stolons at or near the surface. Leaves are basal, linear (1–6 cm long, 1–2 mm wide), 1-veined, round in cross-section, widest at middle, and tapering to a sharp or blunt tip. Flowers minute (ca. 3 mm long), unisexual (on same plant) with four greenish white petals. One male flower is borne on an erect stalk (3.5 cm long) and 2–4 stalkless, female flowers are obscurely clustered at base of plant. Fruit is small (1–2 mm long), oblong, and nut-like.

Habitat & Distribution
Littorella americana is a native, northern plant of shallow (<1.5 m) lakes. It prefers acidic, low nutrient, softwaters. Uncommon in Vt.

Notes: A colony-forming species, with interconnected horizontal stolons and rhizomes, that grows as an emergent if stranded on receding shores. It can only flower when plants are emersed from water. Often present with other rosette-formers like *Lobelia dortmanna*, *Isoëtes,* and *Eriocaulon.* Similar to American awlwort (*Subularia aquatica*).

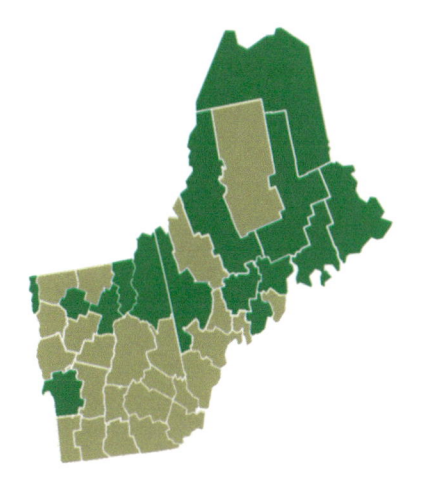

Lobelia — Campanulaceae
Lobelia

WATER LOBELIA — *Lobelia dortmanna* L.

Water lobelia is a short, rooted, submersed, perennial plant. Leaves are clustered in basal rosettes, 2–5 cm long, linear, tubular (two hollow tubes in cross section), blunt-tipped, and arching outward. Flowers (1–2 cm long) have white to bluish, united petals, borne loosely towards the end of an emergent stalk. Flower stalks protrude from water surface about 30 cm (but can be 2–3 m long overall). Fruit a green, barrel-shaped, nodding, capsule.

HABITAT & DISTRIBUTION

Lobelia dortmanna is a native species of still, shallow waters of lake and pond shores. It is widespread in northern New England. This slow-growing plant prefers soft, clear, nutrient-poor waters with a sandy substrate. It is intolerant of high-nutrient conditions.

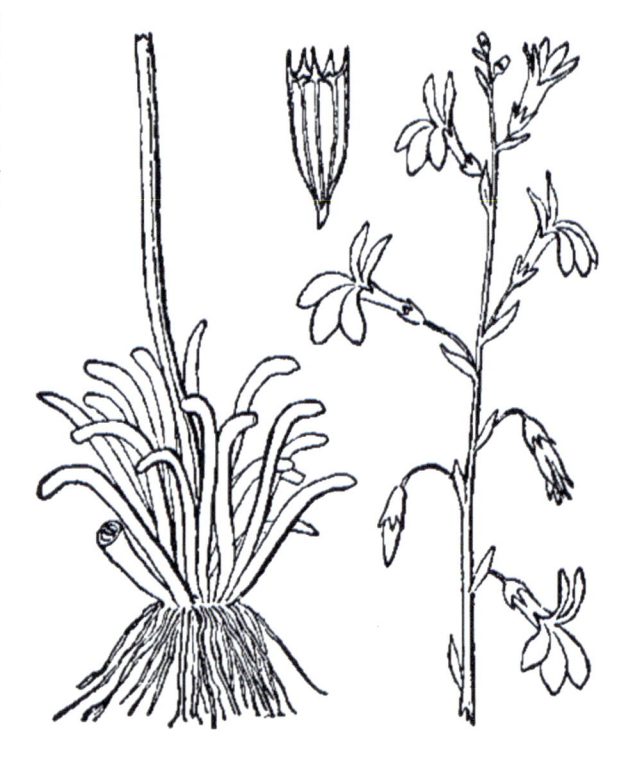

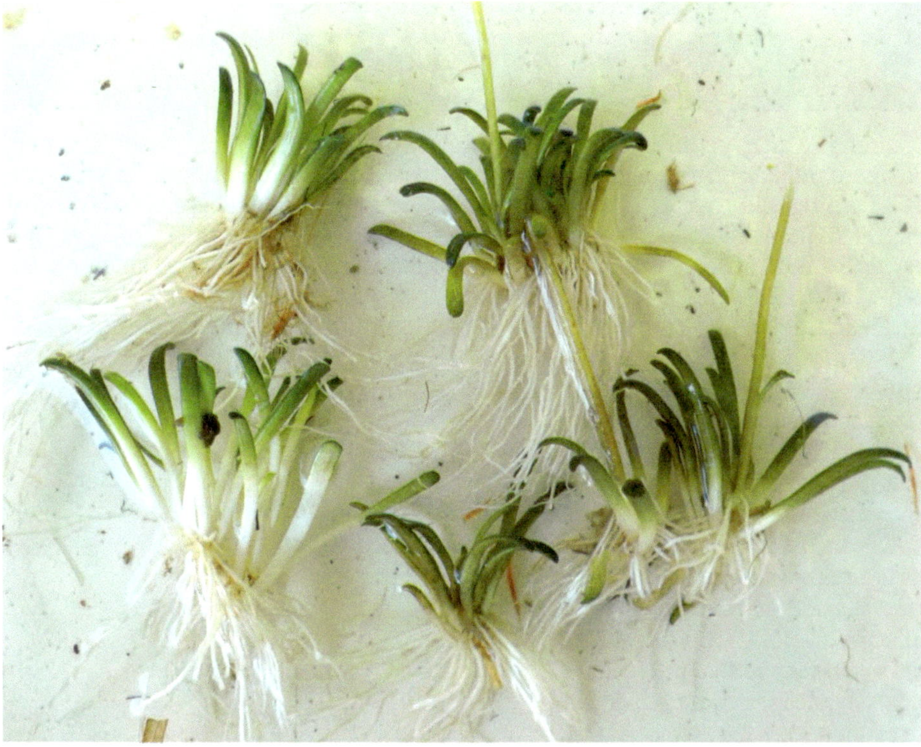

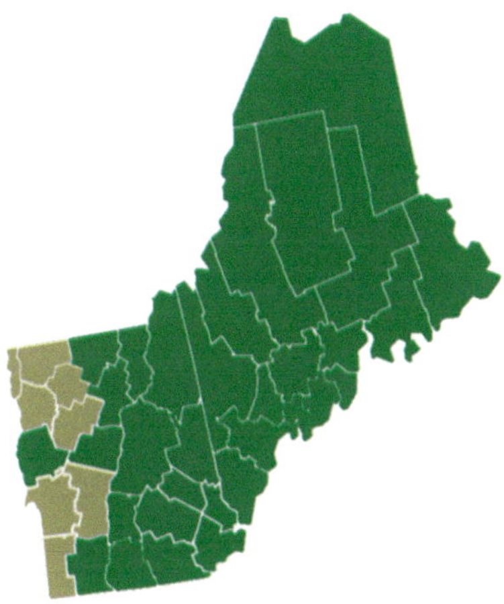

Ludwigia — Onagraceae
Water-primrose

COMMON WATER-PRIMROSE — *Ludwigia palustris* (L.) Ell.

Common water-primrose, or Water purslane, is a mostly submersed (or emergent on mud) perennial, with reddish, many-branched stems (3–50 cm long) laying on sides or partly floating and rooting at lower nodes. Leaves are opposite, green to reddish-bronze, abruptly stalked (5–25 mm long) with egg-shaped to oval, toothless, blades (5–30 mm long, to 20 mm wide). Flowers are inconspicuous, bisexual, borne singly in leaf axils, green-brown, bell-shaped, and topped with 4 triangular sepals. Fruits are oblong (2–5 mm long, 2–3 mm wide), 4-sided capsules with 4 longitudinal green bands.

HABITAT & DISTRIBUTION

Ludwigia palustris is a native species occupying waters of marshes, swamps, rivers and ponds. It is common and widespread in northern New England. Plants can form dense mats in shallow waters.

Note: Similar to the endangered Toothcup [*Rotala ramosior* (L.) Koehne] of sandy coastal pond shores in NH. Toothcup, a more emergent species, however, has somewhat 4-sided stems, more lance-shaped leaves, and small white petals.

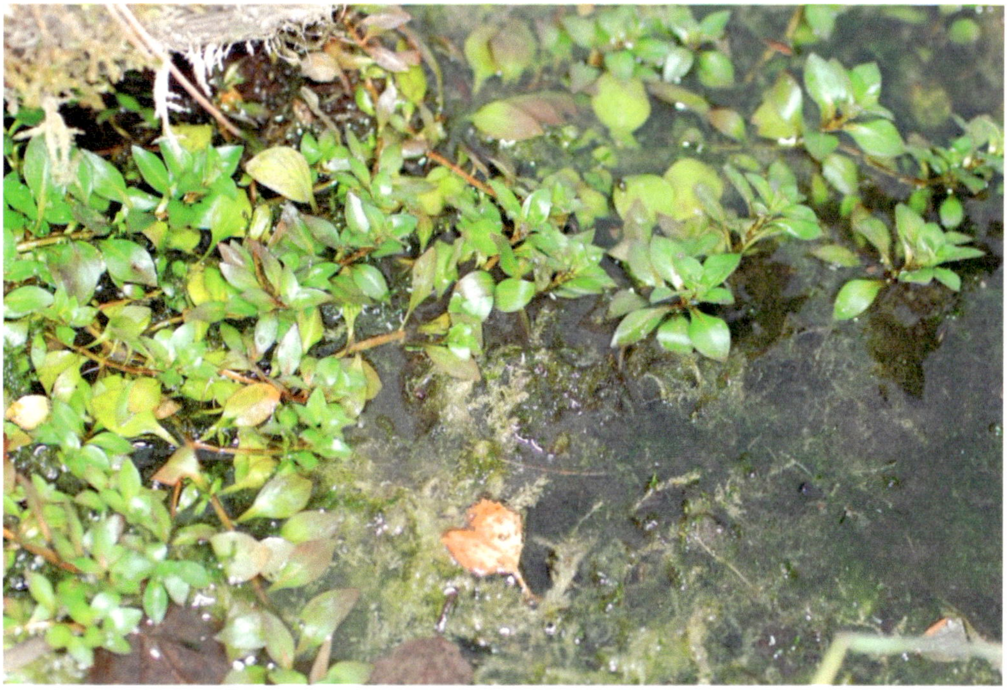

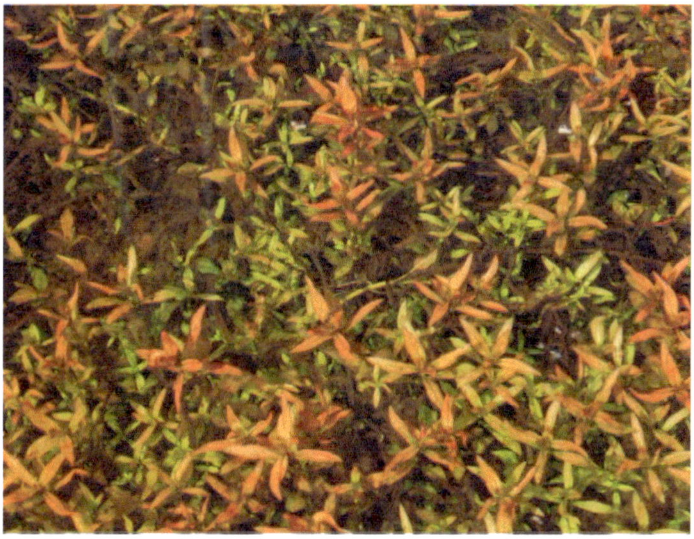

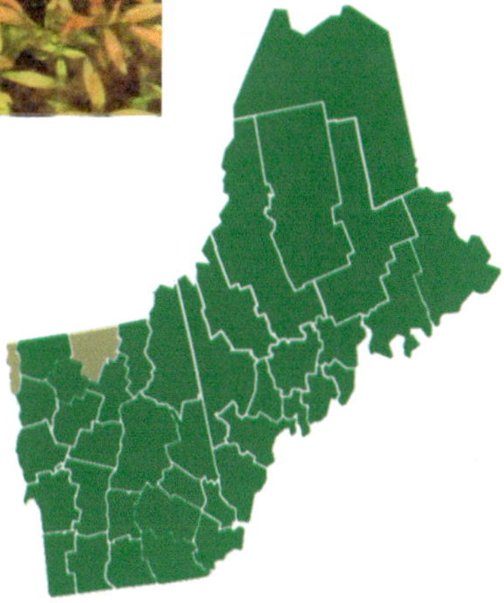

Marsilea — Marsileaceae
Water-clover

EUROPEAN WATER-CLOVER — *Marsilea quadrifolia* L.

European water-clover, or Water-shamrock, is a perennial, aquatic fern that grows from long, creeping rhizomes, rooting mostly at nodes. Leaves are usually floating, with blades centrally divided into four triangular segments (each 7–21 mm long, 6–19 mm wide). Leaf stalks are narrow, flexible, and long (6–16 cm). Reproduces by spores, borne in hairy, pea-shaped (5 mm) structures (sporocarps) produced near base of leaf stalks.

HABITAT & DISTRIBUTION

Marsilea quadrifolia is a non-native plant naturalized in shallow pond or slow rivers. It is uncommon in ME. It was *introduced* into New England (via CT) from Europe in the mid-1800s.

Myriophyllum — Haloragaceae
Water-milfoil

The Water-milfoils are slender, submersed, rooted, perennials with upright stems growing from rhizomes. Leaves are opposite, alternate, or whorled, and mostly feather-like, delicate and finely divided into thread-like segments. The emersed, bract-like, leaves associated with flowers, if present, usually differ in form and are reduced in size. Flowers are unisexual (on same plant), minute, and inconspicuous, borne singly either in leaf axils of submersed leaves or clustered on upright spikes above water. Flowering spikes have male flowers in uppermost positions and female flowers lowermost. Fruits are tiny (to 2.5 mm long), nut-like clusters of four 1-seeded segments.

Eight species occur in northern New England, including two non-natives. A few species are aggressive and highly invasive. Most species are highly variable so reproductive structures and existence of winterbuds (turions) are often essential for proper identification. Winterbuds are formed in late summer and fall as highly condensed clusters of green, apical leaves.

KEY TO SPECIES

1. Leaves (if present) simple, reduced to minute scales (<1 mm long) or bumps, on stiffly upright stems (2–20 cm); floral bracts toothless……………………………….................*M. tenellum*
1. Leaves well-developed, pinnately divided into narrow segments, on elongated flexuous stems; floral bracts toothless, toothed, or deeply divided..2
    2. Leaves alternate, or alternate and whorled on same stem; male flowers with four stamens, 0.5–1 mm long; flowers in axils of submersed leaves………………………….…3
    2. All leaves whorled; male flowers commonly with 8 stamens (4 in *M. heterophyllum*), 1–2 mm long, flowers in emersed spikes... 4
  3. Leaves alternate (sometimes near-opposite); fruit 0.7–1.3 mm long, smooth without ridges; no turions present..*M. humile*
  3. Leaves both alternate and whorled on same stem; fruit 2–2.5 mm long, warty, with prominent ridges; turions present at stem tips in autumn……......................*M. farwellii*
    4. Upper flowers of spike and floral bracts alternate; spikes 2–5 cm tall; leaves with 3-7 pairs of narrow segments………………………………...……………………..... *M. alterniflorum*
    4. Upper flowers of spike whorled; spikes 4–15+ cm tall; leaves with 5–20 pairs of narrow segments……………………………………………………………………………………………… 5
  5. Floral bracts inconspicuous, bracts under lower flowers <2 times as long as the flowers or fruits; bracts under upper flowers with toothless to small-toothed edges…….....…………... 6
  5. Floral bracts conspicuous, bracts under lower flowers ≥2 times as long as the flowers or fruits; bracts under upper flowers with deeply divided to sharp-toothed edges………....…..7
    6. Middle leaves with 5–12 segments on each side; uppermost leaves rounded at tip; cylindrical turions formed in autumn; stems sparingly branched and vertical, stem diameter uniform throughout; non-flowering stem tips usually green and knob-shaped….
    ………………………………………………………………………………………....…. *M. sibiricum*

6. Middle leaves with 13–20 segments on each side; many uppermost leaves flat-topped at end (as if cut off); turions never formed; stems much-branched, often laying on water surface, and thickened (1.5-2 times as thick as lower stem) just below flowering spikes; non-flowering stem tips usually red and tassel-like......................... *M. spicatum*
7. Bracts under female (i.e., lower) flowers sharply toothed; fruits 1–1.5 mm long; turions not widened at tip, formed at base of stems or rhizomes in fall; stems often reddish.......... .... *M. heterophyllum*
7. Bracts under female flowers pinnately lobed or deeply divided; fruits 2–2.5 mm long; turions club-shaped and widened at tip, formed on stems in fall; stems olive green........... .... *M. verticillatum*

ALTERNATE-FLOWERED WATER-MILFOIL *Myriophyllum alterniflorum* DC.

Alternate-flowered water-milfoil is a submersed, densely leaved plant with elongated, flexuous, much-branched stems (0.3–1 mm thick). Leaves are whorled (3–5 per whorl), pinnately divided into 3–7 pairs of narrow segments, and short (3–12 mm long, 3–14 mm wide). Flowers are borne in short (2–5 cm tall) upright spikes, emersed above water, where the upper flowers and floral bracts are alternate (sometimes the lowest are opposite). Floral bracts are short (no more than twice length of flower) and toothless or minutely serrated. Male flowers commonly have 8 stamens (1–2 mm long). Turions never produced.

HABITAT & DISTRIBUTION

Myriophyllum alterniflorum is a northern, native species of mostly shallow, nutrient-poor, lakes and slow streams. It is scattered in northern reaches but most common in ME.

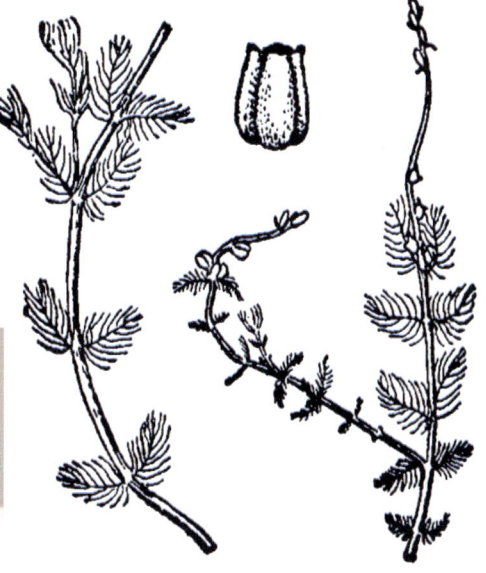

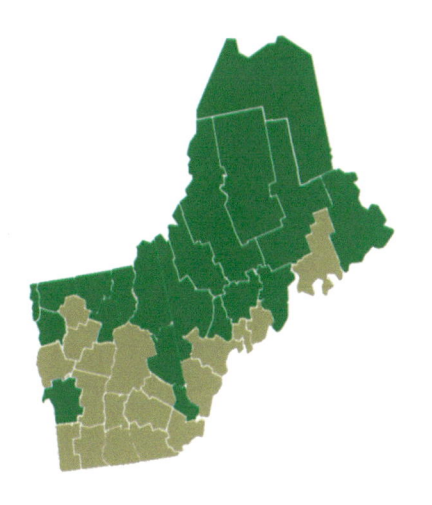

FARWELL'S WATER-MILFOIL — *Myriophyllum farwellii* Morong

Farwell's water-milfoil is a delicate, wholly submersed, densely leaved plant with narrow, flexuous stems. Leaves are both alternate and partly whorled on same stem, 10–25 mm long, and pinnately divided into 5–12 pairs of narrow segments. Flowers and fruits submersed, borne in axils of submersed leaves. Male flowers commonly have 4 stamens (0.5–1 mm long). Fruit (2–2.5 mm long) rough, with prominent, bumpy, vertical ridges. Small, stiff turions produced in autumn at the tips of submersed stems.

HABITAT & DISTRIBUTION

Myriophyllum farwellii is a northern native species of cold, acidic, shallow lakes. It is uncommon in NH and VT.

Notes: Similar to Low water-milfoil (*M. humile*) which produces smooth fruits and lacks turions. The warty fruits of *M. farwellii* are present late July to September.

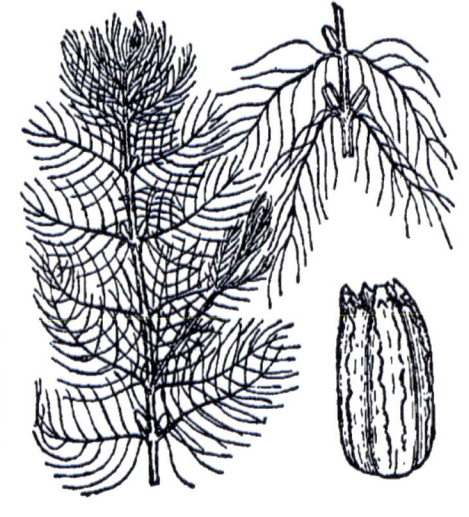

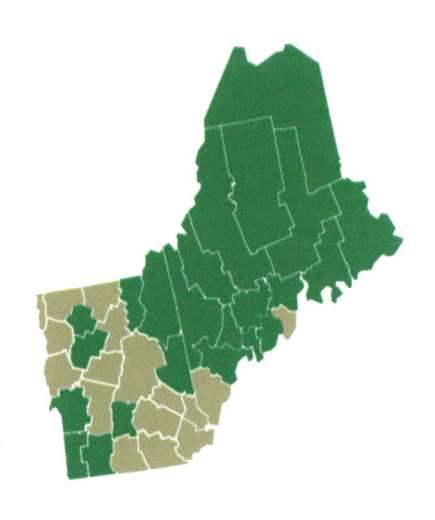

VARIABLE-LEAVED WATER-MILFOIL *Myriophyllum heterophyllum* Michx.

Variable-leaved water-milfoil is a submersed, densely leaved plant with relatively stout, reddish, elongated, much-branched stems, swollen just below the flowering spikes. Submersed leaves (2–4 cm long, 1–3 cm wide) are whorled (4–6 per whorl), closely spaced (whorls <10 mm apart), and pinnately divided into 6–14 pairs of thread-like segments. Flowers are borne in long (15–30 cm tall), leafy, upright, spikes, emersed above water. Floral bracts are whorled, conspicuous, leaf-like, oval (4–18 mm long, much-longer than flowers or fruits), with sharp-toothed edges (bracts near waterline may be somewhat lobed). Male flowers commonly have 4 stamens. Fruits (1–1.5 mm long) are rounded. Turions are produced in fall, at base of stems or rhizomes.

HABITAT & DISTRIBUTION

Myriophyllum heterophyllum is a non-native species of lakes, ponds, and streams. Native to eastern U.S., but this plant is *invasive* and prohibited in New England. Plants can grow rapidly, aggressively spread by stem fragmentation, and develop dense beds which hinder waterbody uses.

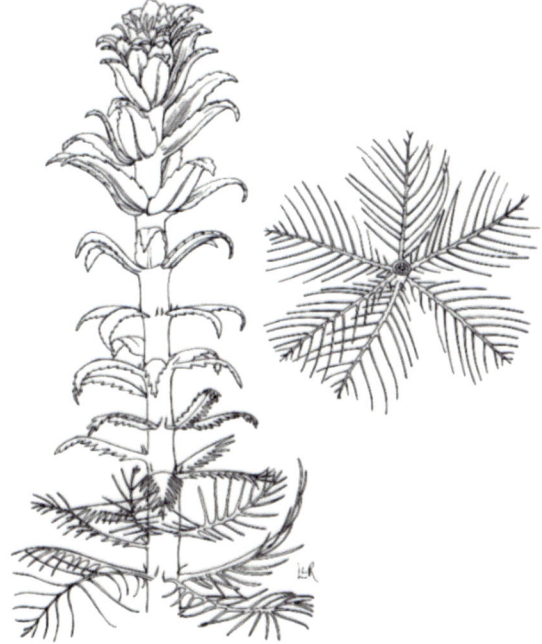

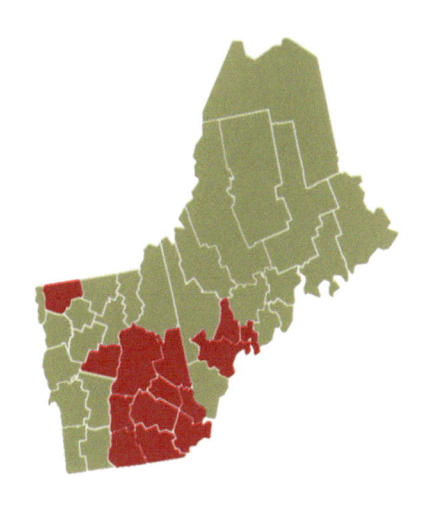

88

LOW WATER WATER-MILFOIL — *Myriophyllum humile* (Raf.) Morong

Low water-milfoil is a delicate, normally submersed, densely leaved plant with elongated, flexuous stems. Leaves are alternate (sometimes some near-opposite), 10–25 mm long, and pinnately divided into 5–12 pairs of narrow segments. Leaves and stems may be brownish red. Flowers and fruits submersed, borne in axils of submersed leaves. Male flowers commonly have 4 stamens (0.5–1 mm long). Fruit (0.7–1.3 mm long) is smooth-sided. No turions produced.

HABITAT & DISTRIBUTION

Myriophyllum humile is a native species in shallow, nutrient poor, mostly acidic, waters of ponds, lakes, and streams. It is rare in Vt.

Notes: This northern species is highly variable in form with stem length and leaf number increasing with water depth. When stranded on receding pond shores, a rigid, simple-leaved growth form can occur.

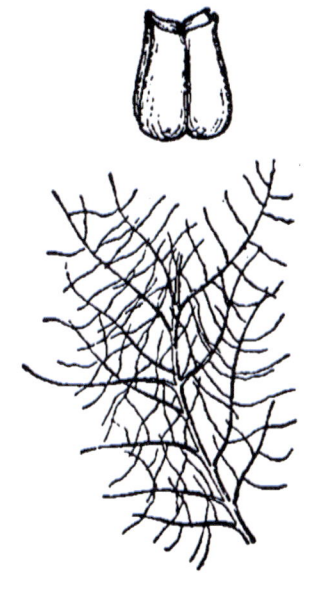

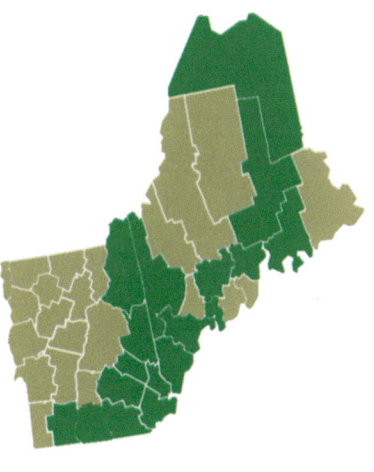

NORTHERN WATER-MILFOIL — *Myriophyllum sibiricum* Komarov

Northern water-milfoil is a submersed, densely leaved plant with slender, few-branched, mostly vertical stems. Tips of non-flowering stems are usually green (except during turion formation) and knob-like. Submersed leaves (1.5–3 cm long) are whorled (3–4 per whorl, whorls ≤10 mm apart), and with well-developed middle leaves, pinnately divided into 5–12 pairs of thread-like segments. Many uppermost leaves rounded at tips. Flowers are borne in upright (4–10 cm tall) spikes, emersed above water. Floral bracts are whorled and inconspicuous. Upper bracts are shorter than flowers with toothless to small-toothed edges and lower bracts <2 times as long as the flowers or fruits. Male flowers have 8 stamens. Dark green, cylindrical (≥2 cm long) turions are produced at stem tips in autumn.

HABITAT & DISTRIBUTION

Myriophyllum sibiricum is a native species of primarily clear, alkaline waters. Uncommon.

Note: Similar to Eurasian water-milfoil (*M. spicatum*) but stems essentially unbranched and vertical, fewer leaf segments, and formation of turions in late season.

EURASIAN WATER-MILFOIL *Myriophyllum spicatum* L.

Eurasian water-milfoil is a submersed, rhizome-less, densely leaved plant with long (1 m+), slender, much-branched stems, swollen (nearly twice as thick) just below the flowering spikes, and curved to lie parallel to water surface. Tips of non-flowering branches are usually reddish and tassel-like. Submersed leaves (3–3.5 cm long, 1–2.5 cm wide) are whorled (3–4 per whorl), widely spaced (whorls 10–30 mm apart), distinctly stalked, and with well-developed middle leaves, pinnately divided into 13–20 (mostly 14–17) pairs of thread-like segments, and flat (segments all in one plane). Many uppermost leaves flat-topped across at end (as if cut off with scissors). Flowers are borne in upright (6–10 cm tall) spikes, emersed above water. Floral bracts are whorled and inconspicuous. Upper bracts are shorter than flowers with toothless to small-toothed edges and lower bracts <2 times as long as the flowers or fruits. Male flowers have 8 stamens. Turions are never produced.

HABITAT & DISTRIBUTION

Myriophyllum spicatum is a non-native species established in lakes and ponds. It is **invasive** and prohibited in New England. Occurrences are scattered, primarily in more alkaline waters.

Notes: This species, a renown aggressive weed, exhibits broad environmental tolerances (e.g., depth, pH, temperature, salinity, exposure, nutrient load, turbidity), rapid growth, and easy dispersal via stem fragmentation. It is similar to non-flowering Northern water-milfoil (*M. sibiricum*) plants but has more leaf segments, reddish, tassel-like stem tips, and lacks turions.

SLENDER WATER-MILFOIL — *Myriophyllum tenellum* Bigel.

Slender water-milfoil is a slender submersed plant with short (2–20 cm tall), stiff, upright, unbranched stems arising singly from buried rhizomes. Leaves are alternate, inconspicuous, stubby, reduced to minute scales (<1 mm long), and few, or absent. Flowers are borne in the upper leaf axils of wiry, emergent stems. Floral bracts are alternate, toothless, and about the length of the flowers. Fruits are about 1 mm long. Turions never produced.

HABITAT & DISTRIBUTION

Myriophyllum tenellum is a native species of shallow, mostly acidic, ponds and lakes. It is widespread.

Notes: This northern species is unlike other water-milfoils in New England as plants are essentially leafless. Sterile plants are usually abundant, and populations can be dense, propagating largely by vegetative means. Flowering is infrequent and occurs only in very shallow (e.g., 5 cm deep) water where upright stems may reach 30 cm.

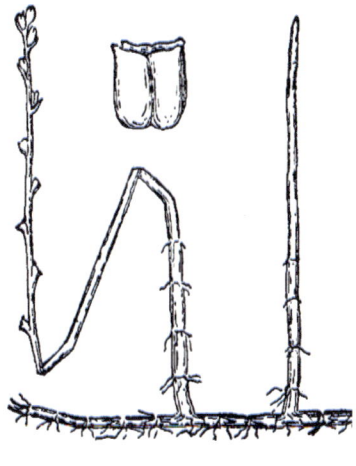

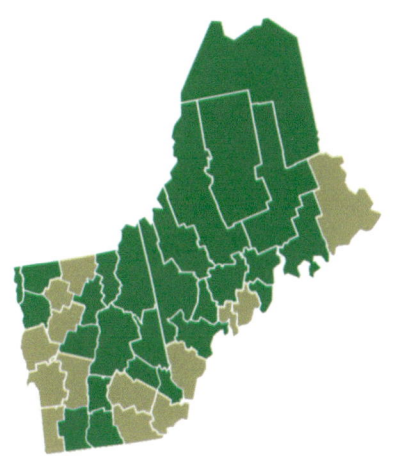

WHORLED WATER-MILFOIL *Myriophyllum verticillatum* L.

Whorled water-milfoil is a submersed, densely leaved plant with elongated, olive green, sparingly-branched stems. Submersed leaves (2–4.5 cm long, 1.5–2.5 cm wide) are whorled (4–5 per whorl) and pinnately divided into 9-17 pairs of thread-like segments. Flowers are borne in long (7–20 cm tall), leafy, upright, spikes, emersed above water. Floral bracts are whorled, conspicuous (to 20 mm long), lower bracts ≥2 times as long as the flowers or fruits, with mostly deeply divided or pinnately lobed edges. Male flowers commonly have 8 stamens. Fruits (2–2.5 mm long) are rounded. Dark, yellow green, distinctively club-shaped (wider at tip than at base) turions (1–3 cm long) are formed on stem tips in fall.

HABITAT & DISTRIBUTION

Myriophyllum verticillatum is a native species of lakes and streams, with an affinity for alkaline waters. It is more common in northern areas.

Najas — Hydrocharitaceae
Waternymph

The Waternymphs, or Naiads, are slender, rooted, much-branched, submersed annuals. Their leaves are opposite (or appearing whorled when extra leaves present), toothed along margins, with linear (<2 mm wide) blades that widen to a stem-sheathing base. Flowers are unisexual (on same plant), submersed, and borne 1–3 per leaf axil. Fruits are small (2–4 mm long), with a thin, membranous covering, one-seeded, and tapered at tip.

Five species occur in northern New England, one being non-native. Fruits and foliage serve as important waterfowl food.

KEY TO SPECIES

1. Leaf margins with 20–100 minute teeth per side (visible with hand lens); top of basal leaf sheath tapered or sloping..2
1. Leaf margins with 7–17 teeth per side; top of basal leaf sheath distinctly lobed or abruptly widened..4
   2. Leaves tapered from about the middle, to long, slender, pointed tips; seeds smooth and glossy, widest above the middle, deep brown to yellow......................................3
   2. Leaves abruptly pointed, tapered only from last 2–3 mm, or rounded at tip; seeds minutely pitted and dull, widest at middle, yellowish white with purple tinge, greenish, or reddish brown.. *N. guadalupensis*
3. Seeds broad (seed length:width ratios <3)...*N. flexilis*
3. Seeds thin and elongate (seed length:width ratios 3–4).................................*N. canadensis*
   4. Leaves generally stiff and recurved in age, with 7–15 course teeth per side (visible without magnification), top of leaf sheath lobed and finely toothed; seeds slightly curved..*N. minor*
   4. Leaves spreading and lax in age, with 13–17 fine teeth per side (visible with magnification), top of leaf sheath abruptly widened, but not lobed, and coarsely jagged; seeds straight... *N. gracillima*

N. flexilis

N. minor

CANADIAN WATERNYMPH — *Najas candensis* Michx.

Canadian waternymph is completely submersed and possesses the same vegetative morphology as its sister species *N. flexilis* (see below). Seeds smooth and shining, length:width ratios 3–4.

HABITAT & DISTRIBUTION

Najas canadensis is a native species of lakes and rivers. Its range overlaps that of its sister species Northern waternymph (*N. flexilis*), where it can occupy the same bodies of water. It is documented from ME, NH, and VT.

Note: Work by Les et al. (2015) has uncovered this cryptic species, which is remarkably similar, yet genetically distinct, from *N. flexilis*. It can be separated by its subtly different "thinner" elongate seeds (i.e., seed length:width ratios >3).

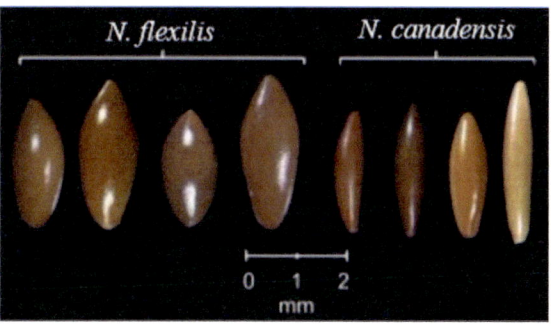

Modified from
D.H. Les et al. 2015. Mol. Phylogen. Evol. 82: 15–30

NORTHERN WATERNYMPH — *Najas flexilis* (Willd.) Rostk. & Schnidt

Northern waternymph is completely submersed, 25–50 cm tall, with much-branched stems. Leaves are 1.5–4 cm long, 0.5–1 mm wide (above widened base), tapered to a slender tip, with 35–80 fine teeth per side (visible with hand lens). Top of basal leaf sheath is gradually tapered. Usually 1 fruit per leaf axil. Seeds smooth and shining, deep brown to yellow, and widest above the middle, length:width ratios <3.

HABITAT & DISTRIBUTION

Najas flexilis is a native species of lakes, ponds, and streams. It is common and widespread throughout northern New England.

Notes: Les et al. (2015) revealed this species is nearly indistinguishable from the cryptic *N. canadensis*. In general, both species have smooth shiny seeds, but the diploid *N. flexilis* has longer and broader seeds (i.e., thicker seeds = seed length:width ratios <3) compared to the thinner seeds of *N. canadensis* (tetraploid). Without seeds, it is difficult to distinguish from Southern waternymph too.

98

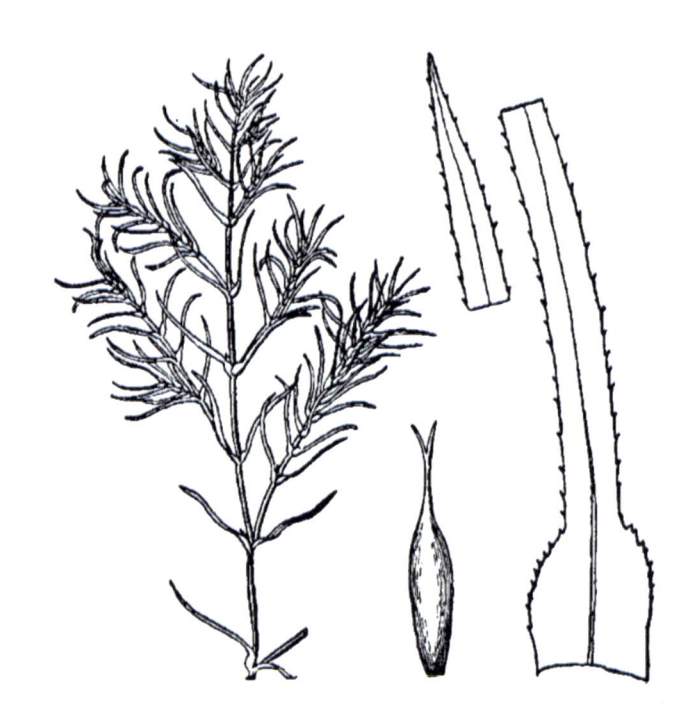

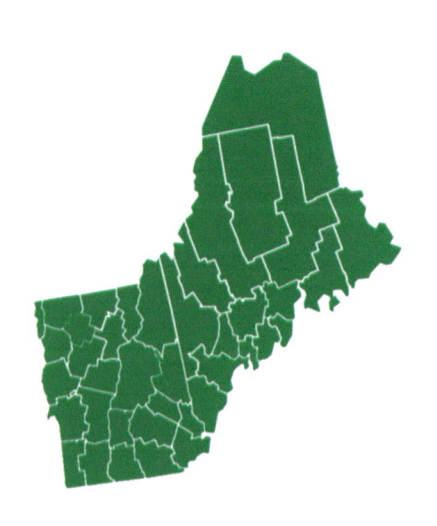

SLENDER WATERNYMPH — *Najas gracillima* (A. Braun) Magnus

Slender waternymph is completely submersed, 4–48 cm tall, with slightly branched, slender (to 0.7 mm wide) stems. Leaves are 0.6–2.8 cm long, 0.1–0.5 mm wide (above widened base), pointed at tip, with 13–17 fine teeth per side (visible only with magnification). Top of basal leaf sheath is abruptly widened, but not lobed, and coarsely jagged. Fruits with style positioned off to one side of apex. Seeds straight, dull, light brown.

HABITAT & DISTRIBUTION

Najas gracillima is a native species of lakes and streams, most abundant in acidic waters. It is uncommon in northern New England. Most similar vegetatively to Northern waternymph (*N. flexilis*) however its leaves have fewer teeth and lack a tapered leaf sheath.

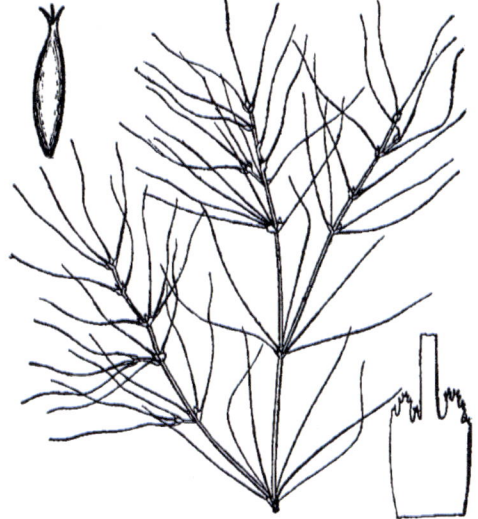

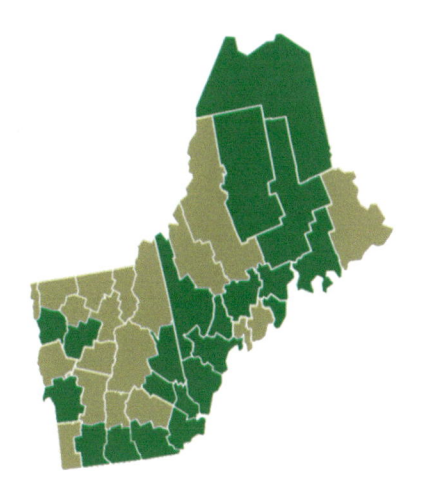

SOUTHERN WATERNYMPH *Najas guadalupensis* (Spreng.) Magnus

Southern waternymph, or Guppy grass, is completely submersed, 10–90 cm tall, with much-branched stems. Leaves are 0.5–3 cm long, 0.2–2.1 mm wide (above widened base), abruptly pointed or rounded at tip, with 50–100 fine teeth per side (visible only with magnification). Top of basal leaf sheath is gradually tapered. Seeds minutely pitted and dull, yellowish white with purple tinge, greenish, or reddish brown, and widest at middle.

HABITAT & DISTRIBUTION

Najas guadalupensis is a native species of alkaline, fresh, or brackish lakes, reservoirs, ponds, and rivers. It is very uncommon.

Note: Two variants of this species exist in New England: ours are the finer-stemmed (0.1–0.8 mm wide) plants with 50–100 teeth on each leaf margin (ssp. *guadalupensis*). Those with more robust stems (1–2 mm wide) and 20–40 teeth (ssp. *olivacea* (Rosendahl & Butters) Haynes & Hellquist) also occur in southern New England.

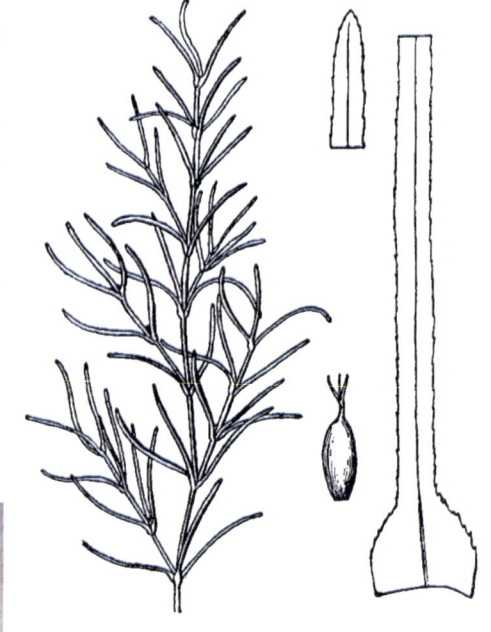

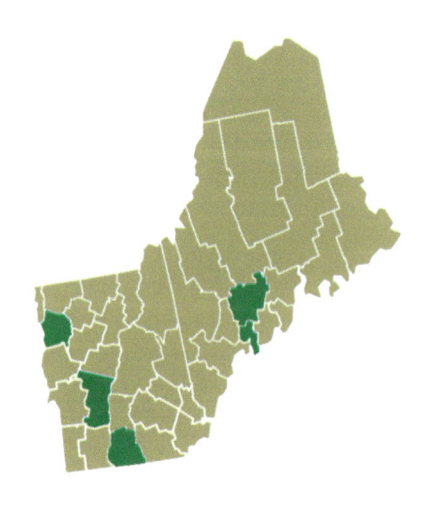

BRITTLE WATERNYMPH *Najas minor* All.

Brittle waternymph, or Brittle naiad, is completely submersed, 11–120 cm tall, compact, with heavily branched stems. Leaves are stiff and recurved in age, 0.5–3.4 cm long, 0.1–1.2 mm wide (above widened base), pointed at tip, with 7–15 course teeth per side (visible without magnification). Top of basal leaf sheath is lobed and finely toothed. Seeds are slightly curved, dull, and purplish brown.

HABITAT & DISTRIBUTION

Najas minor is a non-native species naturalized in lakes and slow rivers. It is restricted mostly to southern areas of states where it is **invasive** and banned. Native to Eurasia, the species can be a problematic weed. Plants are easily spread by stem fragmentation and highly tolerant of turbidity and nutrient-rich waters.

Nelumbo — Nelumbonaceae
Lotus

AMERICAN LOTUS — *Nelumbo lutea* Willd.

American lotus, or Water lotus, is a rooted, perennial with long, slender, horizontal rhizomes and massive, round leaves and flowers. Leaves are circular, without a sinus, large (to 60 cm), floating (flat) or emergent (concave), blue green, with centrally radiating veins, and long (to 2 m), centrally attached stalks. Flowers are very large (10–20 cm wide), bisexual, pale yellow, borne singly, above water, on long stalks (to 2 m). Flowers have a central, green, cone-shaped, flat-topped, receptacle. Fruits are 1-seeded, nut-like (10–16 mm), and embedded in a leathery receptacle.

HABITAT & DISTRIBUTION

Nelumbo lutea is a non-native species occurring in still, usually shallow (<1 m), waters of lakes, ponds, or tidal marshes, in full sun. It is limited to southern ME, where it was **introduced**.

Notes: This distinctive plant is cultivated as an ornamental in water gardens. It is fast-growing and can become aggressive. It possesses water-repellant (hydrophobic) leaves and flowers which generate heat to attract pollinators.

Nuphar — Nymphaeaceae
Pond lily

The Pond lilies, or Spatterdocks, are rooted, perennials with broad, leathery, exposed leaves (mostly floating, sometimes emersed) and submersed membranous leaves. Foliage is generally heart-shaped, with rounded basal lobes, and a defined mid-vein/pinnate venation. Flowers are showy, spherical, and bright yellow, emersed singly atop a stout, rigid stalk. Flowers are bisexual with 5–6 petal-like sepals, inconspicuous scale-like petals, and numerous strap-like stamens surrounding a central, flattened stigmatic disk. Fruits are green to purplish, leathery, and berry-like.

Three species, all native, occur in northern New England. The name *Nuphar lutea* (L.) Sm., a strictly Eurasian species, is often misapplied to Pond lilies of North America.

KEY TO SPECIES

1. Exposed leaf blades large (13–36 cm long), leaf stalks 4–12 mm wide; flowers large (2–4.5 cm wide) with 6 sepals and a yellow disk; rhizomes stout (2.5–7 cm diam.)......................2
1. Exposed leaf blades small (4–13 cm long); flowers small (1–2 cm wide) with 5 sepals and a dark red, deeply-lobed disk; rhizomes slender (0.5–2.5 cm diam.) ....*N. microphylla*
    2. Exposed leaves floating, leaf stalks flattened and laterally winged; inside surface of sepals purplish...*N. variegata*
    2. Exposed leaves emersed, sometimes floating, leaf stalks round in cross-section; inside surface of sepals usually green... *N. advena*

IMMIGRANT POND LILY *Nuphar advena* (Ait.) Ait. f.

Immigrant pond lily is a robust plant with stout rhizomes. Exposed leaves are usually emersed (sometimes floating), large (12–40 cm long) and broad (10–31 cm wide), green, with stout (4–11 mm wide) rounded stalks. Flowers (2–4.5 cm wide) have 6 sepals (inside surface green), long anthers (3–9 mm) and a yellow disk. Fruits are barrel-shaped (1.6–5 cm wide), green, and strongly vertically ribbed.

HABITAT & DISTRIBUTION

Nuphar advena is native to intertidal waters of southern coastal Maine. It is rare in ME where it is state-listed as *special concern*.

Notes: This species is very common in southeastern US. It reaches its northern limits in New England. Populations in ME represent disjunct occurrences, with closest populations possibly being in southeastern NY.

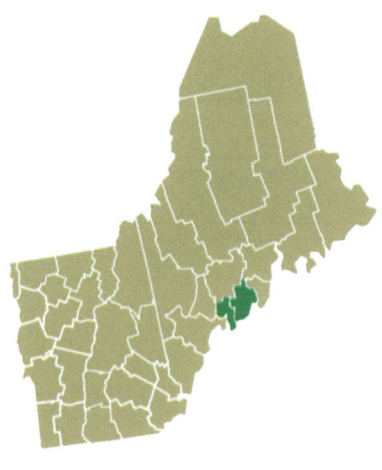

SMALL-LEAVED POND LILY   *Nuphar microphylla* (Pers.) Fern.

Small-leaved pond lily, or Tiny cow lily, has floating leaves arising from slender, submersed rhizomes. Floating leaves (4–13 cm long, 3–8 cm wide), green to purple, typically have a deep sinus (up to half the blade length) with weak, compressed stalks (1–2.5 mm wide). Flowers (1–2 cm wide) have 5 sepals, short anthers (1–3 mm) and a dark red, deeply-lobed disk. Fruits are urn-shaped (1–2 cm wide), with an obvious neck and smooth wall.

HABITAT & DISTRIBUTION

Nuphar microphylla is a native species in lakes, ponds, and slow rivers. Uncommon and scattered in northern New England. It is rare in NH, where it is state-listed as *endangered*, but considered historic.

Notes: This diminutive, quite distinctive, pond lily is imperiled throughout much of its range. In areas where it overlaps with *N. variegata*, the two species naturally hybridize to yield plants—with largely intermediate features and limited fertility—recognized as *N. ×rubrodisca* Morong. The hybrid is common in the Lake Champlain Valley of VT.

105

YELLOW POND LILY — *Nuphar variegata* Dur.

Yellow pond lily, or Spatterdock, is a is a robust plant with stout rhizomes. Floating leaves are large (13–36 cm long) and broad (10–24 cm wide), green to purplish, with broad (4–10 mm wide), flattened, stalks that usually winged along their length. Flowers (2.5–4.5 cm wide) have 6 sepals (inside surface purple-tinged), long anthers (3–11 mm) and a yellow disk. Fruits are barrel-shaped (1.5–4 cm wide), green to purple, and strongly vertically ribbed.

HABITAT & DISTRIBUTION

Nuphar variegata is native to lakes, ponds, bogs, deep marshes, ditches, and slow-moving rivers.

It is widespread, common, and usually abundant throughout the region.

Notes: This species is ecologically tolerant and can colonize deeper (to 2 m) waters. Its long, stout, branched rhizomes can be uprooted by storms and beavers and subsequently float. It is known to naturally hybridize with *N. microphylla* to yield plants—with largely intermediate features and limited fertility—recognized as *N. ×rubrodisca* Morong. The hybrid is common in the Lake Champlain Valley of V⊤.

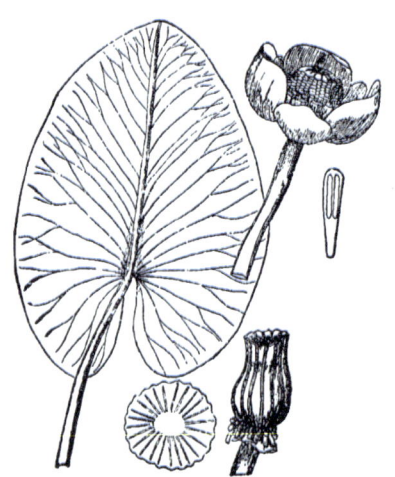

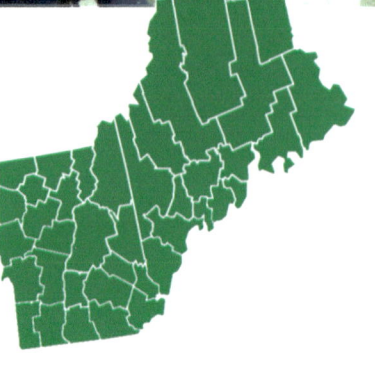

Nymphaea — Nymphaeaceae
Water lily

The Water lilies are rooted, perennials with broad, leathery, floating blades with with long, flexible leaf stalks and pointed basal lobes. Foliage is green above, usually purple beneath, with centrally radiating veins. Flowers are showy, white to pinkish, emersed just at the water surface (as if floating) and borne singly atop stout stalks. Flowers are bisexual with numerous petals and stamens surrounding a central, yellow stigmatic disk. Fruits are green to purplish, leathery, and berry-like.

Two species, both native, occur in northern New England. Water lilies are an important group of ornamental plants cultivated for water gardens. As such, occasionally cultivated varieties are introduced into local waters.

KEY TO SPECIES

1. Flowers 6–19 cm wide with 17–43 petals and 40–100 stamens; leaves 5–40 cm wide, round, with narrow sinus; rhizome horizontal and branched .... *N. odorata*
1. Flowers 3–7.5 cm wide with 8–15 petals and 20–40 stamens; leaves 2–15 cm wide, elliptic, with wide sinus; rhizome vertical and unbranched .... *N. leibergii*

DWARF WATER LILY *Nymphaea leibergii* Morong

Dwarf water lily, or Pygmy water lily, has erect, unbranched, rhizomes. Leaves are broadly elliptical in shape (3–19 cm long, 2–15 cm wide) with a fairly open sinus (lobes greatly divergent). Flowers are conspicuous (3–7.5 cm wide), with yellow centers and 8–15 white petals.

HABITAT & DISTRIBUTION

Nymphaea leibergii is a native boreal species of clear, cold, shallow ponds and lakes where it reaches its southern limit in northern New England. It is rare and state-listed as *threatened* in ME and *endangered* in VT.

Notes: Superficially like a much smaller version of *N. odorata*, yet this diminutive species is also differentiated by presence of submersed foliage and flowers lacking fragrance. Blooming lasts for 5–7 days with flowers opening in afternoon and closing by early evening. It is suspected to naturally hybridize with *N. odorata* to produce sterile plants with features intermediate between the presumed parents.

FRAGRANT WATER LILY — *Nymphaea odorata* Ait.

Fragrant water lily has thick, horizontal, branching rhizomes. Leaves are generally circular in shape (8–40 cm wide), with a closed or narrow sinus. Flowers are large (6–19 cm wide), showy, fragrant, with yellow centers and numerous (17+), white to pale pink, petals. Fruits are globose (4 cm), leathery, and berry-like, maturing underwater on coiled stalks.

HABITAT & DISTRIBUTION

Nymphaea odorata is a conspicuous, native species of still or slow waters of lakes, bogs, marshes, swamps, and rivers. It is widespread and common in northern New England. It has broad tolerances of water depth (0.5–3 m), nutrients, and pH levels. Flowers are insect-pollinated and exhibit diurnal blooming (opening in early morning and closing late afternoon) over a 3-day period.

Notes: It is represented by two subspecies in northern New England. Unlike typical plants, ssp. *tuberosa* (Paine) Wiersema & Hellquist is more robust with rhizomes forming numerous, detachable tubers. Its leaves have green leaf stalks with purplish stripes and blades mostly green underneath. This uncommon variant is considered native to calcareous waters of Lake Champlain Valley but introduced and scattered elsewhere.

subsp. *odorata*

subsp. *tuberosa*

subsp. *tuberosa*

Nymphoides — Menyanthaceae
Floating-heart

The Floating-hearts are rooted, perennials of quiet waters with shiny, heart-shaped floating leaves which superficially resemble those of juvenile waterlilies (*Nuphar* or *Nymphaea*). Plants have slender, anchored rhizomes and long (to 2 m), buoyant, flexuous stems that bear leaves and flower clusters at their summit. Flowers are showy, with 5 united petals, emersed or floating, and borne in clusters from the base of the leafstalk—appearing as if borne on a long leaf stalk. Fruits are elliptical, green capsules.

Two species occur in northern New England, one is non-native. Both species are highly ornamental and cultivated for the water garden trade.

KEY TO SPECIES

1. Flowers white, 5–10 mm wide; flowering stems with a single leaf (3–5 cm long) at base of flower cluster; clusters of submersed spur-like roots present at the leaf base, near water surface... *N. cordata*
1. Flowers yellow, 20–25 mm wide; flowering stems with a pair of leaves (each 5–15 cm long) at base of flower cluster; usually lacking spur-like roots at base of leaves............ *N. peltata*

LITTLE FLOATING-HEART *Nymphoides cordata* (Ell.) Fern.

Little floating-heart is a submersed plant with floating leaves produced singly from the base of a flower cluster. The floating blades are heart-shaped (3–5 cm long), with rounded lobes, and usually mottled with purple. Submersed clusters of slender (2–4 mm thick), fleshy roots are produced at the leaf base, near water surface. Flowers (5–10 mm wide) are white with yellow centers, bloom above water, with 5 fused petals. Fruits are small (4–5 mm long) capsules.

HABITAT & DISTRIBUTION

Nymphoides cordata is a native species in lakes, ponds, and slow streams. It is widespread throughout most of northern New England.

Notes: Plants prefer shallow waters (<1 m) and are tolerant of either nutrient-poor or -rich systems. The clusters of tuber-like roots with attached floating leaves can detach as plantlets for dispersal purposes.

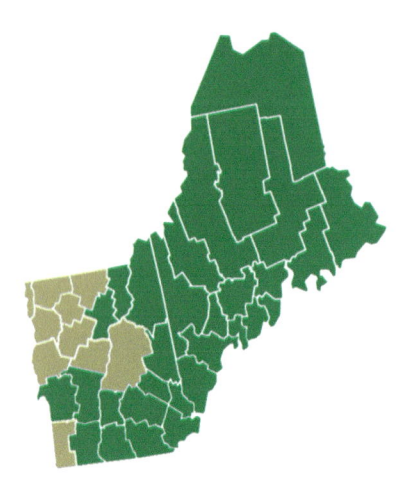

YELLOW FLOATING-HEART — *Nymphoides peltata* (Gmel.) Ktze.

Yellow floating-heart is a submersed plant with floating leaves produced in pairs from the base of a flower cluster. The floating blades are heart-shaped (5–15 cm long), with rounded lobes, and wavy margins. Flowers (20–25 mm wide) are wholly bright yellow, bloom above water, with 5 fused petals. Fruits (12–25 mm long) are beaked, flattened, capsules containing seeds edged with stiff hairs.

HABITAT & DISTRIBUTION

Nymphoides peltata is a non-native species of lakes, ponds, and slow rivers. It is uncommon in ME and Vt. This Eurasian native—introduced for ornamental purposes—is **invasive** and prohibited in New England.

Notes: The first recorded U.S. occurrence of this plant was in 1882 in Winchester, MA. Plants are generally stout and highly productive, most common in nutrient-rich waters, and tolerant of waters to 4 m deep.

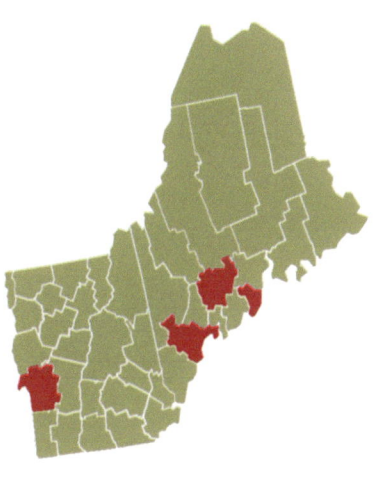

Persicaria — Polygonaceae
Smartweed

WATER SMARTWEED — *Persicaria amphibia* (L.) Gray

Water smartweed is a rooted, floating-leaved perennial with rhizomes that grow prostrate, and sprawling, branched stems. Leaves are alternate, floating, hairy to hairless, short-stalked, with elliptical blades (2–15 cm long, to 17–29 mm wide) and pointed or rounded tips. There is a cylindrical stipule (an ocrea) sheathing the stem at each node. Flowers are small (4–6 mm long), pink, 5-lobed, bisexual, and borne in an erect, dense, many-flowered spike-like cluster (10–30 mm long) that terminates the stem above water. Fruits are small (2–3 mm), flattened, dark brown, shiny, achenes.

HABITAT & DISTRIBUTION
Persicaria amphibia is native species of shallow, still or slow-moving waters in marshes, lakes, and rivers. It is scattered in northern New England.

Notes: A very variable species, typically growing underwater or floating, but even emersed if water levels drop. It is similar to the more widespread emergent wetland species, Scarlet smartweed [*P. coccinea* (Muhl. ex Willd.) Greene], which can grow floating leaves if it becomes inundated. It has much wider leaves (32–63 mm) and longer (30–90 mm) flower clusters.

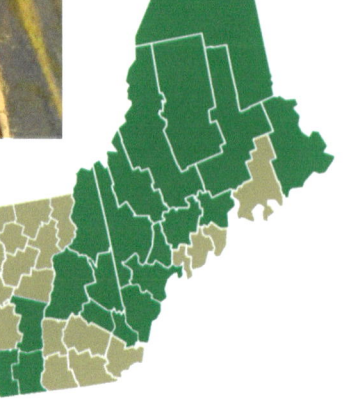

Podostemum — Podostemaceae
Riverweed

HORN-LEAVED RIVERWEED　　*Podostemum ceratophyllum* Michx.

Horn-leaved riverweed, or Threadfoot, is an olive-green, algal-like, submersed, perennial with flattened, leathery stems, and creeping, green roots adhering to hard substrates via fleshy disks. Leaves are alternate, long (2–142 mm), and repeatedly forked into linear or hair-like (0.05–0.8 mm wide) segments. Flowers are inconspicuous, water-pollinated, bisexual, and borne singly along stems. Fruits are small (1.5–3 mm long), 2-valved capsules.

HABITAT & DISTRIBUTION
Podostemum ceratophyllum is a native species of rocky, fast-flowing rivers and streams of good water quality. It is uncommon, but probably overlooked. Usually growing in areas of rapids, attached to ledge, rocks, or boulders.

Note: This is a highly unusual flowering plant—often mistaken for mosses or algae—occupying a highly specialized habitat. Plants become reproductive as water levels drop exposing them to air.

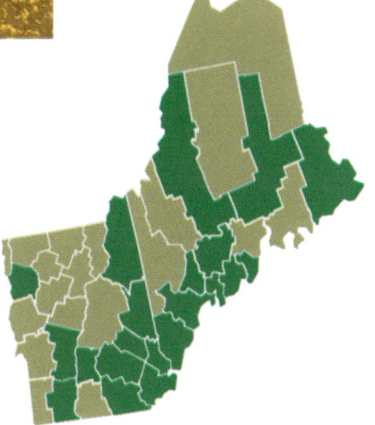

Pontederia — Pontederiaceae
Pickerelweed

PICKERELWEED — *Pontederia cordata* L.

Pickerelweed is a submersed (or emersed) perennial with rhizomes rooted in mud. Leaves are submersed in a basal rosette. Blades are without stalks, linear, thin, ribbon-like (6–25 cm long, 4–6 mm wide), and with an observable midvein. Plants can sometimes have narrow emersed leaves.

HABITAT & DISTRIBUTION
Pontederia cordata is a native species of lake shores, ditches, and brackish tidal marshes. It is common throughout northern New England. Plants are very variable in shape of leaves.

Notes: The submersed life-form (f. *taenia* Fassett) of the species is described here. Typically, the species exists as the larger, emergent, life-form. The emergent form, very common on muddy shores, is more robust overall and has stout, long-stalked (to 60 cm), glossy, heart-shaped leaf blades (to 12 cm wide) and dense spikes of showy, blue-purple flowers borne on upright stems. The submersed life-form is similar to Bur-reed (*Sparganium*) and Arrowhead (*Sagittaria*).

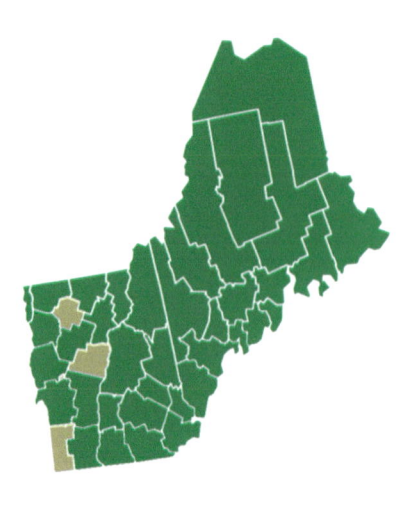

Potamogeton — Potamogetonaceae
Pondweed

The Pondweeds are mostly submersed, rooted perennials with simple or branched stems, regularly with slender, creeping rhizomes. Leaves are alternate (uppermost may appear opposite), mid-veined, and submersed and or floating. All species have submersed leaves, which are thin, ribbon-like or broader, and usually flaccid. Some species additionally have floating blades which are firmer, and elliptical in shape. At the base of each leaf, there is an open (margins unfused) or closed (fused) stipule that forms a cylindrical sheath around the stem. Sometimes a portion of the stipule is fused to base of leaf. Flowers are small, individually inconspicuous, and clustered mostly into narrow, cylindrical, wind-pollinated spikes held above water (or in submersed, few-flowered clusters). Fruits are small, hard, seed-like, achenes. Winter-buds (turions) are produced by some species at tips of branches or in leaf axils. Winter buds are firm, much-condensed stems with crowded stipules and reduced, thickened leaves.

Twenty-seven species occur in northern New England, including 12 rare, and one introduced species. Pondweeds comprise the most diverse group of aquatic plants in our region and collectively have high ecological importance as wildlife food. Identification is sometimes difficult and may require mature fruits. Vegetative features are highly variable and influenced by environmental conditions (water depth and flow).

Pondweeds have a propensity for natural interspecific hybridization which can confound identification. Research supports plants long referred to as Ogden's pondweed (*P. ogdenii* Hellq. & Hilton)—listed as rare in alkaline waters of Vermont—are natural hybrids. Like other *Potamogeton* hybrids in our area, details of these plants are omitted here.

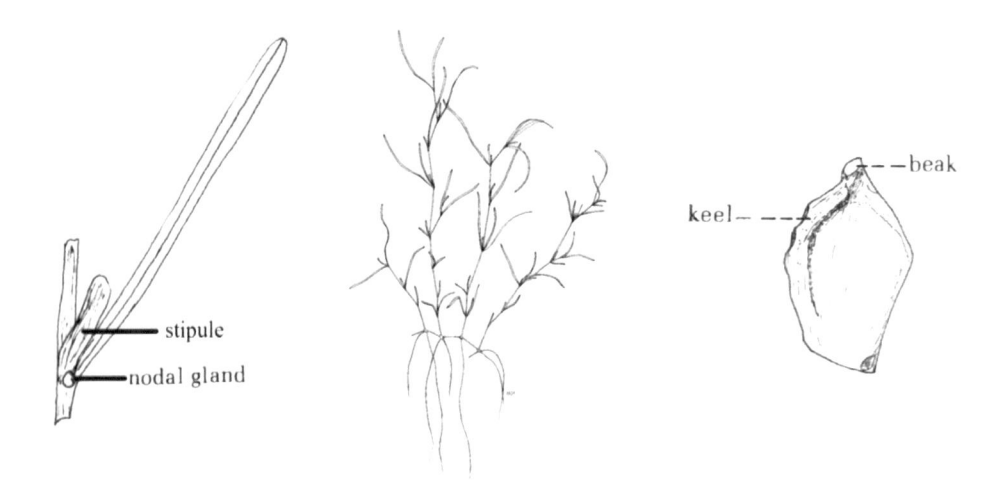

KEY TO SPECIES

1. Stipule fused to base of submersed leaves for a portion its length, remaining tip of stipule extending as a ligule.. 2
1. Stipule completely free from leaf base...4
    2. Leaves conspicuously two-ranked so that whole plant is flat; leaf blades stiff, with finely toothed edges, 20–60 veins, and a pair of round lobes at base; plants without floating leaves.. *P. robbinsii*
    2. Leaves not conspicuously two-ranked; blades lax, with smooth edges, <20 veins, and not lobed at base; plants with or without floating leaves...3
3. Submersed leaf blades 0.5–2 mm wide, blunt-tipped; floating leaf tips rounded; fruit ca. 1.5–2.5 mm in diameter (including keel), keel blunt and poorly developed, beak absent.. *P. spirillus*
3. Submersed leaf blades 0.1–0.4 mm wide, pointed or long-tapering at tip; floating leaf tips acute; fruit ca. 1–1.5 mm in diameter, keel well-developed, beak present or absent..*P. bicupulatus*
    4. Plants with floating leaves.. 5
    4. Plants without floating leaves...14
5. Floating leaf blades 6–15 mm long, 5–9 veined; submersed leaves 0.1–1 mm wide, 1–3 veined; flower spikes 3–8 mm long; fruit 1.5–2.5 mm wide................................ *P. vaseyi*
5. Floating leaf blades 15–120 mm long, many-veined (7–49); submersed leaves 0.25–10 mm wide, 1–37 veined; flower spikes 10–50 mm long; fruit 2.0–4.5 mm wide................6
    6. Submersed leaves linear, flattened or phyllodial (appearing as a bladeless leafstalk) 7
    6. Submersed leaves broadly linear-oblong, lance-shaped to elliptical............................ 9
7. Submersed leaves 2–10 mm wide, thin and transparent, 3–13 veined, with prominent lacunae band 1–2 mm wide on each side of mid-vein; fruits with distinct, acute keels on back and sides..*P. epihydrus*
7. Submersed leaves 0.25–2.5 mm wide, thick and nearly opaque, phyllodial (appearing as a bladeless leaf stalk), 3–5 veined, lacking lacunae bands; fruit without keels on sides........... 8
    8. Floating blades 5–10 cm long, usually rounded or somewhat lobed at base, with 17–35 veins, leaf stalk pale and thin at blade end; stipules 4–10 cm long; fruit 3.5–5 mm wide, wrinkled on sides, keel inconspicuous.. *P. natans*
    8. Floating blades 2.5–6 cm long, rounded to tapering at base, with 7–23 veins, leaf stalk usually not pale at blade end; stipules 2.5–4 cm long; fruit 2–3.5 mm wide, not wrinkled on sides, with a prominent keel..*P. oakesianus*

9. Submersed leaves sessile, without a distinct stalk..10
9. Submersed leaves with a narrow stalk (petiole)..11
   10. Stems branched; submersed leaves green to brownish, tips usually pointed, margins with minute, 1-celled, teeth; fruit laterally compressed...................... *P. gramineus*
   10. Stems unbranched; submersed leaves usually reddish, tips blunt, margins smooth; fruit plump.. *P. alpinus*
11. Floating leaves 19–49 veined; submersed leaf margins undulating (crisped); stems rusty spotted to black-spotted; fruits keeled...12
11. Floating leaves 11–29 veined; submersed leaf margins flat, rarely crisped; stem lacking spots...13
   12. Stems black-spotted; submersed leaves with 7–19 veins, 10–25 mm wide, slightly arching; floating leaf blades with 15–19 veins; fruit 3.1–4.1 mm.............................. *P. pulcher*
   12. Stems rusty-spotted; submersed leaves with 19–49 veins, 25–75 mm wide, distinctly arching; floating leaf blades with 27–49 veins; fruit 3.9–5.2 mm................. *P. amplifolius*
13. Submersed leaves with long stalks (5–15 cm long), blades with 2–5 rows of lacunae on each side of mid-vein, larger leaves with blades 10–35 mm wide; leaf tips pointed; fruit red to red-brown, the lateral keels with blunt to sharp teeth .... *P. nodosus*
13. Submersed leaves stalkless or with short stalks (0.5–4 cm long), larger leaves with blades 20–45 mm wide; leaf tip with an abrupt, short, stiff point; fruit gray-green to olive-green, the lateral keels without teeth .... *P. illinoensis*
   14. Leaf margins finely but distinctly and sharply toothed especially towards tip; fruit with prominent (2–3 mm long), curved, beak; turions commonly formed.............*P. crispus*
   14. Leaf margins without distinct teeth; fruit beak <0.9 mm long; turions absent or rarely formed 15
15. Leaves linear, ribbon-like, thread-like, flattened or phyllodial (appearing as a bladeless leafstalk)...16
15. Leaves broadly linear-oblong, lance-shaped, or elliptical...30
   16. Plants with obvious, elongated rhizomes..17
   16. Plants lacking rhizomes or rhizomes not apparent..20
17. Leaves delicate, thread-like, 0.1–0.5 mm wide, 1-veined; flowering stalks 5–25 cm long.. *P. confervoides*
17. Leaves flattened and ribbon-like or appearing as bladeless stalks, 2–11 mm wide, 3–13 veined; flowering stalks 1.5–8 cm long... 18
   18. Submersed leaves 2–10 mm wide, thin and transparent, 3–13 veined, with prominent lacunae band 1–2 mm wide on each side of mid-vein; fruits with distinct, acute keels on back and sides.. *P. epihydrus*
   18. Submersed leaves 0.25–2 mm wide, thick and nearly opaque, phyllodial (appearing as a bladeless leaf stalk), 3–5 veined, lacking lacunae bands; fruit without keels on sides...... 19
19. Submersed leaves 0.7–2.5 mm wide, 3–5 veins; fruit 3.5–5 mm wide, wrinkled on sides, keel inconspicuous... *P. natans*
19. Submersed leaves 0.25–1 mm wide, 3 veins; fruit 2–3.5 mm wide, not wrinkled on sides, with a prominent keel...*P. oakesianus*
   20. Leaves 1-veined, 0.1–1 mm wide..21
   20. Leaves 3–35 veined, 1.2–5 mm wide..22
21. Circular raised glands present on at least some nodes, green, gold, brown, or rarely white, to 0.5 mm wide; stipules 5–20 mm long; flowering stalks 0.5–6 cm long..........*P. gemmiparus*
21. Nodal glands absent; stipules 4–12 mm long; flowering stalks 0.5–3 cm long.......... *P. vaseyi*
   22. Circular raised glands present on at least some nodes...................................23
   22. Nodal glands absent... 28
23. Stipules fibrous, often whitish or brown...24

23. Stipules delicate, not fibrous, green, brown, or white..26
  24. Leaf blades obtuse or rounded at tip, sometimes with a short, projected tip; leaves 5–7 veined; winter buds with inner leaves oriented at a right angle to the outer leaves, outer leaves corrugated at the base, the inner leaves modified into a fan-shaped structure......... .... *P. friesii*
  24. Leaf blades obtuse to tapering at tip; winter buds flattened, the inner and outer leaves oriented in the same plane, the outer leaves not corrugated at the base, the inner leaves undifferentiated.. 25
25. Stems conspicuously flattened, stiff, 0.6–3.2 mm wide; leaf blades with 15–35 veins; fruit 4–4.5 mm wide; stipules conspicuous, 15–35 mm long, with overlapping margins................ .... *P. zosteriformis*
25. Stems not flattened, flexuous, <1 mm wide; leaf blades with 3–5 veins; fruit 1.9–2.1 mm wide; stipules inconspicuous, 6–16 mm long, with margins fused to form a cylinder or tube around stem... *P. strictifolius*
  26. Leaf blades 1–3.5 mm wide, bluntly pointed or rounded at tip, often reddish; fruit 2.5–3.5 mm wide; winter buds 3.5–7.8 cm long, 2.3–5.1 mm wide...................... *P. obtusifolius*
  26. Leaf blades 0.2–2.5 mm wide, pointed at tip, green; fruit 1.5–2.2 mm wide; winter buds 0.9–3.2 cm long, 0.3–1.8 mm wide 27
27. Stipules with fused margins, forming a tube surrounding the stem; leaf blades with up to 2 rows of lacunae on each side of the midvein; flowering stalks 10–62 mm long, usually terminal, 1–3 per plant; spike usually of 2–4 distinct, interrupted whorls of flowers/fruits...... .... *P. pusillus*
27. Stipules wrapped around the stem, but with unfused, overlapping margins; leaf blades with up to 5 rows of lacunae on each side of the midvein; flowering stalks 5–30 mm long, axillary or terminal, more than 3 per plant; spike usually of 1–3 crowded whorls of flowers/fruits.. *P. berchtoldii*
  28. Stems conspicuously flattened (to 3.2 mm wide); leaf blades with 15–35 veins; spikes cylindrical with 7–11 whorls of flower; flowering stems stout and long (20–50 mm); fruits 4–4.5 mm wide.. *P. zosteriformis*
  28. Stems rounded or slightly compressed (to 1 mm wide); leaf blades with 3–5 veins; spikes knob-like with 1–5 whorls of flowers; flowering stems short (3–14 mm); fruits 1.4–4 mm wide 29
29. Leaves acute at tip; fruit 1.4–2.3 mm wide, with a single, wing-like, bumpy keel; flowering stalks 3–11 mm long and axillary.. *P. foliosus*
29. Leaves bristle-tipped (rarely blunt); fruit 2.3–4 mm wide, with three keels; flowering stalks 6–14 mm long, axillary and terminal... *P. hillii*
  30. Submersed leaf blades with a stalk..31
  30. Submersed leaves without a stalk..34
31. Submersed leaf margins undulating (crisped); stems rusty-spotted to black-spotted........32
31. Submersed leaf margins flat, rarely crisped; stems lacking spots..33
  32. Stems black-spotted; submersed leaves with 7–19 veins, 10–25 mm wide, slightly arching; fruit 3.1–4.1 mm... *P. pulcher*
  32. Stems rusty spotted; submersed leaves with 19–49 veins, 25–75 mm wide, distinctly arching; fruit 3.9–5.2 mm.. *P. amplifolius*
33. Submersed leaves with long stalks (5–15 cm long), larger leaves with blades 10–35 mm wide; leaf tips pointed; fruit red to red-brown, the lateral keels with blunt to sharp teeth..... .... *P. nodosus*
33. Submersed leaves stalkless or with short stalks (0.5–4 cm long), larger leaves with blades 20–45 mm wide; leaf tip with an abrupt, short, stiff point; fruit gray-green to olive-green, the lateral keels without teeth.. *P. illinoensis*

34. Leaf blades not clasping the stem..35
34. Leaf blades with base conspicuously clasping the stem..37
35. Stems unbranched; submersed leaves reddish, tip usually blunt, margins without teeth; fruit plump...*P. alpinus*
35. Stems branched; submersed leaves green to brownish, tip usually pointed, margins with minute, 1-celled, teeth; fruit laterally compressed..36
    36. Leaves with 3–9 veins, with 1–2 rows of lacunae on either side of midvein; fruits 1.9–2.3 mm with beak 0.3–0.5 mm long..*P. gramineus*
    36. Leaves with 7–19 veins, with 2–5 rows of lacunae on either side of midvein; fruits 2.5–3.6 mm with beak 0.5–0.8 mm long.. *P. illinoensis*
37. Stipules conspicuous and persistent; leaf blades often >10 cm long, toothless, the tips boat-shaped (hood-like); rhizomes rusty-red spotted; fruit 4–5.7 mm wide with a prominent keel; flowering stems usually >18 cm long... *P. praelongus*
37. Stipules inconspicuous or soon disintegrating; leaf blades usually <10 cm long, with minute, 1-celled teeth on margins, the tips flat; rhizomes unspotted; fruit 1.6–4.2 mm wide with an inconspicuous keel or without a keel; flowering stems usually <10 cm long...........38
    38. Leaf blades egg-shaped to near circular, rarely to 5 cm long, delicately 7–15 veined; stipules delicate, disintegrating and absent on lower part of stem; fruit 1.6–3 mm wide......
    ....*P. perfoliatus*
    38. Leaf blades narrowly egg-shaped to narrowly lance-shaped, mostly 3–10 cm long, coarsely 13–21 veined; stipules disintegrating into coarse, white fibers; fruit 2.2–4.2 mm wide..*P. richardsonii*

P. perfoliatus

P. gemmiparus

ALPINE PONDWEED

Potamogeton alpinus Balb.

Alpine pondweed, or Red pondweed, is a submersed, rooted, perennial, with round, slender, unbranched stems (to 200 cm long) and creeping rhizomes. Leaves of two types: all submersed or occasionally both submersed and floating. Submersed leaves are alternate, lax, toothless, linear to lance-shaped (4.5–18+ cm long, 5–20 mm wide), without stalks, reddish green, 7–9 veins, with 0–6 rows of lacunae on either side of midvein, and blunt at tip. Stipules are inconspicuous (1.5–2.5 cm long), light brown to reddish, sheathing, and completely free from leaf base, blunt at tip. Floating leaves are stalked (stalks 0.1–1.2 cm long, continuous in color to blade junction), blades reddish green, elliptic to oblong (4–7 cm long, 10–25 mm wide), 9–13 veins, and tapering at base. Flower clusters are dense, cylindrical (1–3.5 cm long) and emersed above water (flowering stem 3–10 cm long, erect, axillary or terminal). Fruit plump, olive brown, 3–3.5 mm long, with a keel on back and possibly on sides, and a curved beak (0.5–0.9 mm). Turions are not produced.

HABITAT & DISTRIBUTION

Potamogeton alpinus is a native northern species of shallow to deep, cold-water pools, ponds, and slow streams. It is most frequent in northern portions of states. Rare and state-listed as *endangered* in NH.

Notes: A distinctive feature of this plant is its often strong reddish color, especially in upper portions and flowering stems. The dark red color becomes more evident when plants are dried.

Similar to Grassy pondweed but differs by having red-tinged, longer (to 18 cm), submersed leaves, unbranched stems, and larger, plump fruits. It is known to naturally hybridize with *P. perfoliatus* in VT.

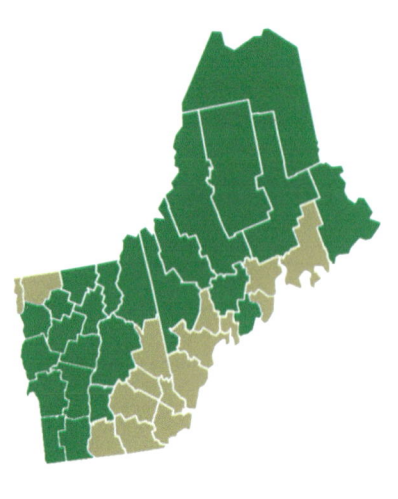

BIG-LEAVED PONDWEED — *Potamogeton amplifolius* Tuck.

Big-leaved pondweed is a robust, submersed, rooted, perennial, with round, often rusty-spotted, few-branched stems (6–110 cm long) and creeping rhizomes. Leaves of two types: all submersed or both submersed and floating. Submersed leaves are alternate, with narrow, black-spotted stalks (0.5–4.5 cm long), lax, distinctly arching, often folded, broadly elliptic (5–12.5 cm long, 2.5–7.5 cm wide), undulating margins, green, often turning reddish-brown, 19–49 veins, lacunae absent, and pointed at tip. Stipules are conspicuous (1.5–11.7 cm long), light brown, sheathing, and completely free from leaf base. Floating leaves are stalked (stalks 2.3–22.6 cm long), blades light green, elliptic to narrowly egg-shaped (4.3–9.2 cm long, 2.5–3.8 cm wide), 27–49 veins, and rounded to lobed at base. Flower clusters are dense, cylindrical (3.4–6.5 cm long) and emersed above water (flowering stem 4.5–22 cm long, erect, axillary or terminal). Fruit reddish brown, 3.9–5.2 mm wide, with a keel on back and sides, and an erect beak (0.5–0.8 mm). Turions are not produced.

HABITAT & DISTRIBUTION

Potamogeton amplifolius is a native species of shallow to deep, lakes, ponds, and rivers. It is widespread in northern New England. An easily recognized pondweed with its large, often reddish, strongly arching, many-veined, submerged leaves, and ability to colonize deeper waters.

Notes: Similar to Spotted pondweed (*P. pulcher*) but with rusty-spotted stems and larger, arching, many-veined, submersed blades. When floating leaves are absent, plants can be confused with Illinois pondweed (*P. illinoensis*). Known to hybridize with *P. illinoensis* (in VT), *P. praelongus* (ME, VT) and *P. pulcher* (NH).

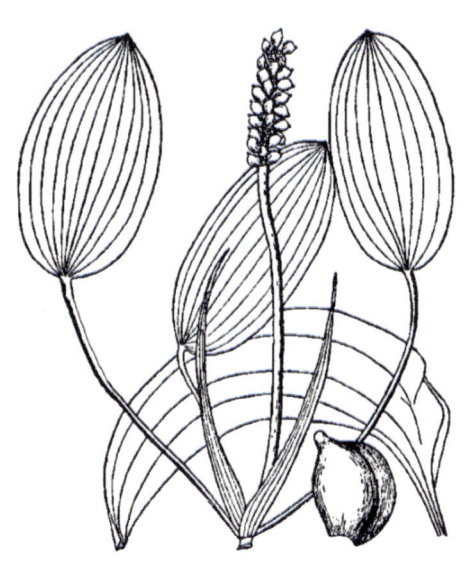

124

BERCHTOLD'S PONDWEED   *Potamogeton berchtoldii* Fieber

Berchtold's pondweed is a delicate, submersed, rooted, perennial, with slender stems (18–150 cm long). Leaves are all submersed, alternate, flaccid, linear (0.9–5.4 cm long, 0.2–2.5 mm wide), pale green to olive-green, 1–5 veins, lacunae present in 1–5 rows on each side of vein, and pointed at tip. Paired, circular, raised glands (–0.5 mm wide), green (same as stem), are inconspicuous at a few nodes. Stipules are inconspicuous, 3.1–9.2 mm long, brown to green or white, delicate, not shredding at tip, wrapped around the stem with overlapping (unfused) margins, and completely free from leaf base. Flower clusters are mostly knob-like (1.5–10 mm long), with 1–3 crowded whorls of flowers/fruits, emersed above water or submersed, at end of cylindric stems (0.5–3.0 cm long), more than 3 per plant. Fruit green to brown, 1.5–2.2 mm wide, not keeled, sides rounded, and with an erect beak (0.1–0.6 mm). Turions are common, slender (0.9–3.2 cm long, 0.3–1.8 mm wide), soft, and leaves two-ranked, with inner leaves modified into a hardened structure.

HABITAT & DISTRIBUTION
Potamogeton berchtoldii is a native species of shallow lakes, pools, and streams. It is widespread and common in northern New England.

Notes: This is our most common narrow-leaved pondweed. It is similar to Small pondweed (*P. pusillus*), but with more (1–5 rows) lacunae per side of leaf midrib, >3 flowering stems per plant, and uninterrupted flowering spikes. Known to hybridize with *P. perfoliatus* (in ME, VT).

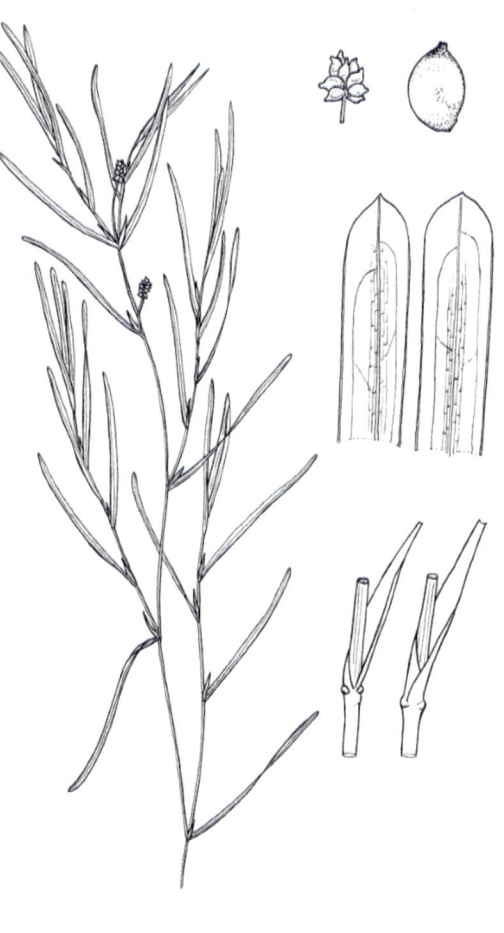

126

SNAIL-SEED PONDWEED — *Potamogeton bicupulatus* Fern.

Snail-seed pondweed is a submersed, rooted, perennial, with compressed, highly branched, slender stems (10–25 cm long) and creeping rhizomes. Leaves of two types: all submersed or both submersed and floating. Submersed leaves alternate, linear and thread-like (1.5–11 cm long, 0.1–0.4 mm wide), light green, 1 vein, lacunae absent, and pointed or long-tapering at tip. Stipules are inconspicuous, light green, fused to base of submersed leaves for <½ of stipule length, with remaining tip of stipule free (usually greater than fused portion). Floating leaves are stalked (stalks 3–35 mm long), blades elliptic to lance-shaped (6–23 mm long, 1–11 mm wide), 3–7 veins, with acute to long-tapering leaf tips. Flower clusters of two types: either globular (1.5–7 mm long) and submersed (flowering stem 1–10 mm long, somewhat recurved, axillary) or short-cylindrical (3–14 mm long) and emersed above water (flowering stem 4–22 mm long, erect or recurved, axillary or terminal). Fruit greenish brown, 1–1.5 mm wide (including keel), with well-developed bumpy keel and usually lacking a beak. Embryo distinctly coiled with more than a full spiral. Turions are not produced.

HABITAT & DISTRIBUTION

Potamogeton bicupulatus is a native, northeastern species of shallow, usually acidic, lakes and streams. It is uncommon in ME and VT.

Notes: This is one of only two species in our flora that produces two types of flowering spikes (emersed and submersed). The other species is Northern snail-seed pondweed (*P. spirillus*) which is similar to *P. bicupulatus* but has ribbon-like submersed leaves, fused portion of stipules about half stipule length, and rounded floating blades.

ALGA PONDWEED
Potamogeton confervoides Rchb.

Alga pondweed is a very delicate submersed, rooted, perennial, with round, much-branched stems (10–80 cm long) and obvious, elongated, creeping rhizomes. Leaves alternate, submersed, thread-like (2–6.5 cm long, 0.1–1 mm wide), flaccid, pale green, 1 vein, and long-tapering and bristly at tip. Stipules are inconspicuous, pale green, and completely free from leaf base. Flower clusters short (5–12 mm long), emersed above water, at end of unusually long stems (5–25 cm long). Fruit light green, 2–3 mm long, 1.7–2.8 mm wide, with keels and erect beak (0.5 mm). Turions (0.7–2 cm long) present in leaf axils of old leaves.

HABITAT & DISTRIBUTION
Potamogeton confervoides is a northern, native species of shallow, acidic, lakes and ponds. It is uncommon throughout.

Notes: This plant is found in our most acidic waters (≤ pH 5) including bog ponds. Its leaves are so delicate and hair-like that when the plant is removed from the water, they collapse onto each other. Easy to overlook if not flowering.

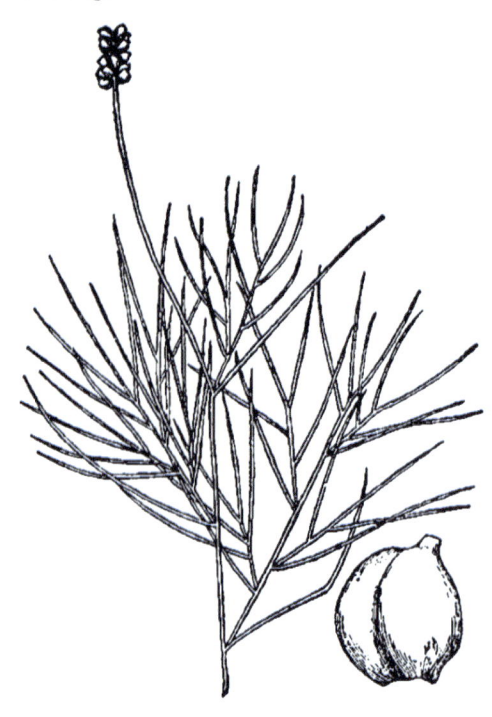

CURLY PONDWEED *Potamogeton crispus* L.

Curly pondweed is a submersed, rooted, perennial, with flattened, few-branched, slender stems. Leaves are all submersed, alternate, bright green, slightly stiff, linear-oblong (12–90 cm long, 4–10 cm wide), round at tip, with 3–5 veins and 2–5 rows of lacunae on each side of midvein. Leaf margins are wavy and conspicuously sharp toothed especially towards tip. Stipules are brown, inconspicuous, and completely free from leaf base. Flowers are borne above water in cylindrical spikes (10–15 mm long) with 6–20 flowers. Fruits (6 mm long, 2.5 mm wide) are brown to red brown, with a prominent (2–3 mm long) curved beak and wavy keel. Turions (1.5–3 cm long, 2 cm wide), consisting of hard, rolled leaves, are commonly formed in leaf axils or at stem tips.

HABITAT & DISTRIBUTION

Potamogeton crispus is a non-native species of alkaline, often polluted, lakes and slow, freshwater to brackish rivers. It is **invasive** and banned in New England.

Notes: Introduced from Europe, this species can be a problematic weed. It is an unusually early blooming species and produces turions in early summer. Plants typically go dormant and decay in late summer as turions germinate into small plants.

RIBBONLEAF PONDWEED *Potamogeton epihydrus* Raf.

Ribbonleaf pondweed is a submersed, rooted, perennial, with slightly flattened, slender stems (10–90 cm long) and creeping rhizomes. Leaves of two types: all submersed or both submersed and floating. Submersed leaves alternate, flat, linear (5–22 cm long, 2–10 mm wide), red-brown to light green, thin and transparent, 3–13 veined, with prominent lacunae bands 1–2 mm wide on each side of mid-vein, and rounded at tip. Stipules are inconspicuous, reddish brown to light green, sheathing, completely free from leaf base. Floating leaves are stalked (stalks 2–13 cm long), blades light green, elliptic (2.5–9 cm long, 4–20 mm wide), 11–41 veins, with rounded leaf tips. Flower clusters are dense, cylindrical (0.8–4 cm long) and emersed above water (flowering stem 1.5–5 cm long, erect, and axillary). Fruit greenish brown, 2.5–4.5 mm wide, with acute keels on back and sides and an erect beak (0.5 mm). Turions are not produced.

HABITAT & DISTRIBUTION

Potamogeton epihydrus is a native species of shallow, acidic or alkaline, lakes, ponds, and slow rivers. It is widespread and common in northern New England.

Notes: Probably our most common pondweed. It is known to hybridize with Clasping-leaved pondweed (*P. perfoliatus*) in ME.

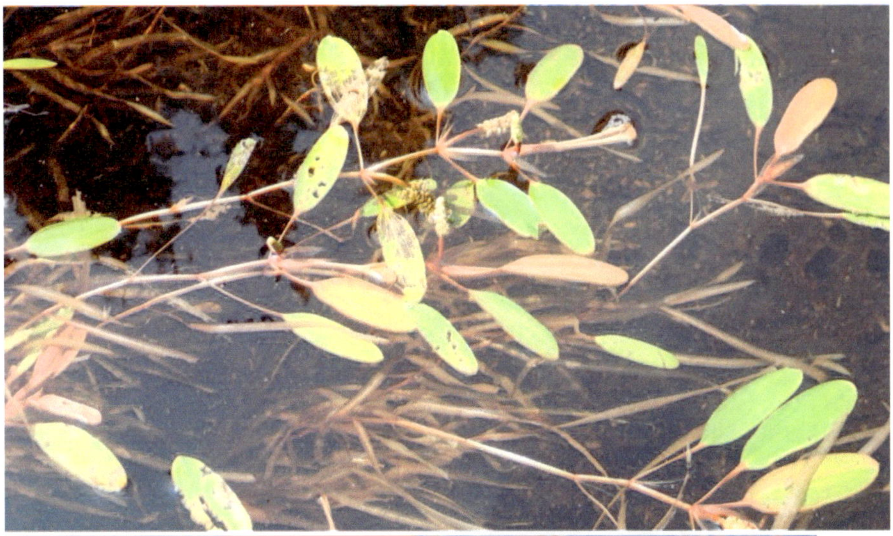

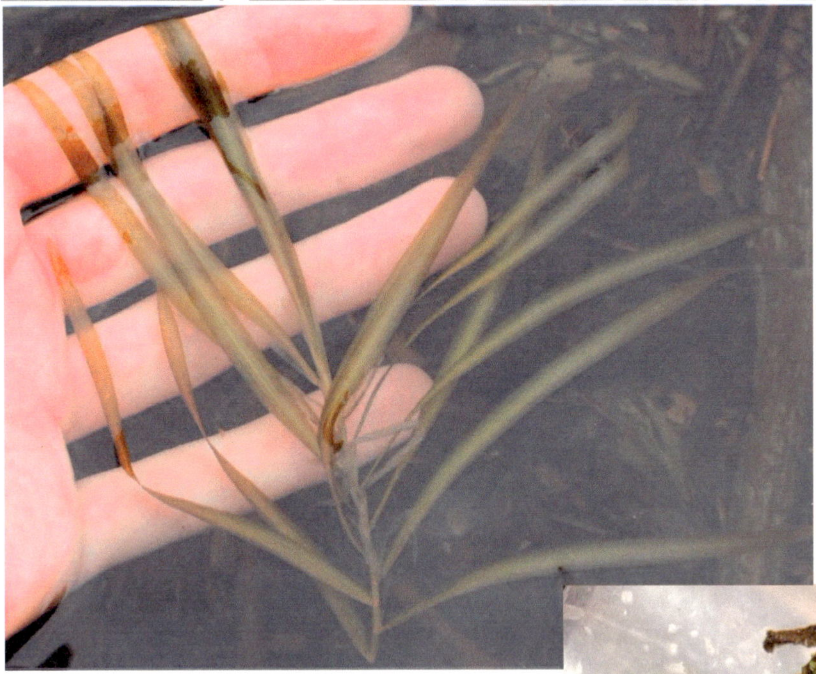

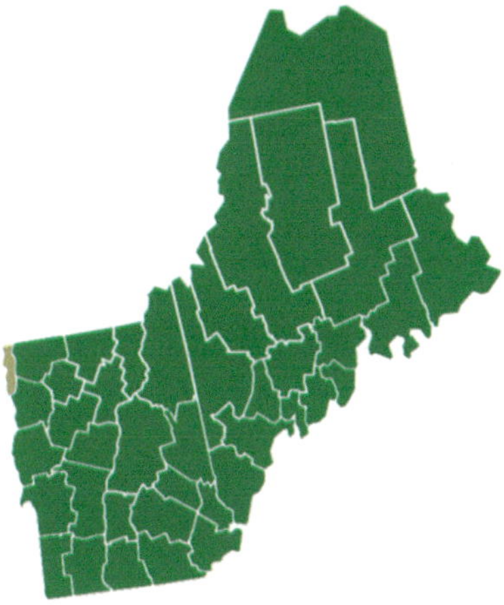

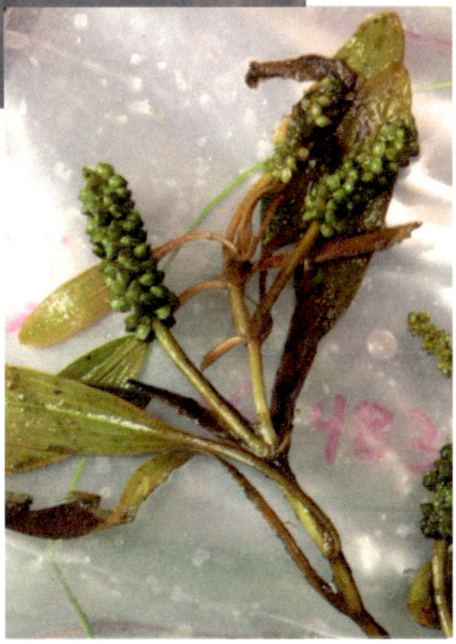

LEAFY PONDWEED *Potamogeton foliosus* Raf.

Leafy pondweed is a delicate, submersed, rooted, perennial, with much-branched, slightly compressed stems (4–75 cm long). Leaves are all submersed, alternate, flaccid, linear (1.3–8.2 cm long, 0.3–2.3 mm wide), pale green to olive-green, 3–5 veins, lacunae rarely present in 0–2 rows on each side of midvein, and pointed at tip (rarely with bristle). Circular glands (to 0.3 mm wide), black to gold, are rarely present at nodes. Stipules are inconspicuous, 2–22 mm long, brown to green, delicate, wrapped around the stem with overlapping margins, and completely free from leaf base. Flower clusters are knob-like to cylindric (1.5–7 mm long), emersed above water, at end of axillary, short (3–11 mm long) stems. Fruit olive-green to brown, 1.4–2.3 mm wide, with a wing-like, undulating keel on back, and erect beak (0.2–0.6 mm). Turions are uncommon, slender (0.9–2.5 cm long, 0.6–2 mm wide), soft, with inner leaves modified into a hardened structure.

HABITAT & DISTRIBUTION

Potamogeton foliosus is a native species of shallow lakes and streams. It is common in VT, but rare and state-listed as *endangered* in NH.

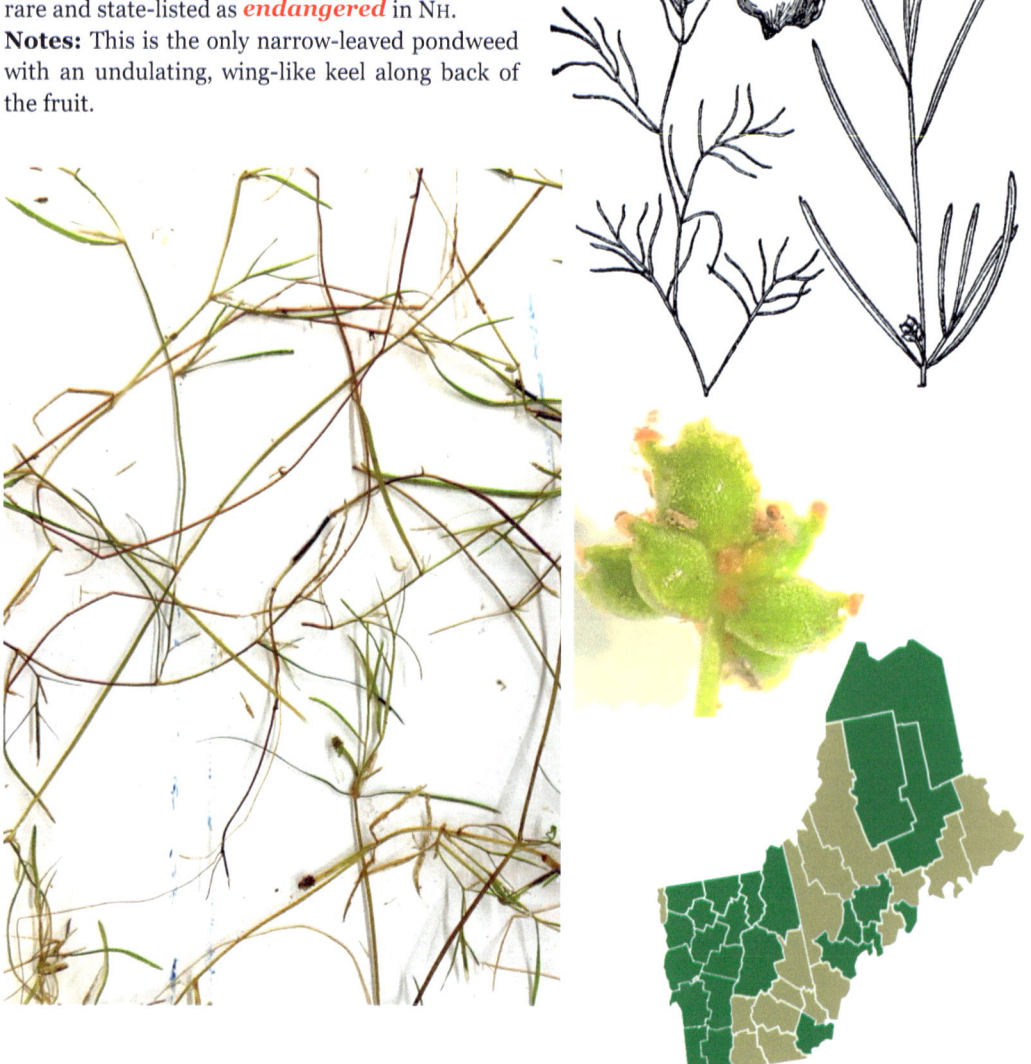

Notes: This is the only narrow-leaved pondweed with an undulating, wing-like keel along back of the fruit.

FLAT-STALKED PONDWEED   *Potamogeton friesii* Rupr.

Flat-stalked, or Fries', pondweed is a delicate submersed, rooted, perennial, with compressed, slender stems (10–135 cm long). Leaves are all submersed, alternate, linear (2.3–6.5 cm long, 1.2–3.2 mm wide), light green, 5–7 veins, lacunae absent or present in 1 row on each side of midvein, and obtuse or rounded at tip, sometimes with a short projected tip. Circular raised glands (to 0.7 mm wide), green, greenish brown, or gold, are usually conspicuous at some nodes. Stipules are inconspicuous, 5–21 mm long, fibrous, white, coarse, shredding at tip, sheathing the stem, and completely free from leaf base. Flower clusters are cylindrical (7–16 mm long), dense, emersed above water, at end of flattened stems (1.2–4 cm long). Fruit olive green to brown, 1.8–2.5 mm long, lacking keels, and with an erect beak (0.3–0.7 mm). Turions are common, slender (15–50 mm long, 1.5–4 mm wide), soft, and leaves four-ranked, with inner leaves modified into a fan-shaped structure oriented at 90° angle to outer leaves, and outer leaves corrugated at base.

HABITAT & DISTRIBUTION

Potamogeton friesii is a native species of shallow, alkaline, lakes and ponds. It is uncommon, rare and state-listed as *endangered*. in ME.

Note: This is the only pondweed with turions where the outer leaves have corrugated bases and the inner leaves turned at right angles to the outer leaves. It is most similar to Straight-leaved pondweed (*P. strictifolius*).

BUDDING PONDWEED — *Potamogeton gemmiparus* (Robb.) JW Robbins

Budding pondweed is a very delicate submersed, rooted, perennial, with round, slender stems (18–150 cm long). Leaves are all submersed, alternate, linear and thread-like (1–6 cm long, 0.2–0.7 mm wide), pale- to olive-green, 1 vein, lacunae absent or present in 0–2 rows on each side of vein, and narrowly tapering to tip. Circular raised glands (to 0.5 mm wide), green, gold, brown, or rarely white, are present on at least some nodes. Stipules are inconspicuous, 5–20 mm long, brown to green or white, sheathing the stem, and completely free from leaf base. Flower clusters are short-cylindrical (1.5–10 mm long), dense and uninterrupted, submersed or emersed above water, at end of stems (1–3.5 cm long). Fruit green to brown, 1.5–2 mm long, lacking keels, and with an erect beak (0.1–0.6 mm). Turions are common as slender (10–30 mm long, 0.3–1.8 mm wide), soft, two-ranked outer leaves and rolled, hardened inner leaves.

HABITAT & DISTRIBUTION

Potamogeton gemmiparus is a native northern species of shallow, acidic waters of lakes, ponds, and slow rivers. It is uncommon, rare and state-listed as *endangered* in NH.

Notes: This narrow-leaved pondweed is geographically restricted to New England and southern Quebec. It is difficult to distinguish from the more widespread Small pondweed (*P. pusillus*) or Berchtold's pondweed (*P. berchtoldii*).

GRASSY PONDWEED — *Potamogeton gramineus* L.

Grassy pondweed is a submersed, rooted, perennial, with round to flattened, slender, branching stems (to 150 cm long) and creeping rhizomes. Leaves of two types: all submersed or both submersed and floating. Submersed leaves are alternate, lax, elliptic (3.1–9.1 cm long, 3–27 mm wide), without stalks, light to brownish green, 3–9 veins, with 1–2 rows of lacunae on either side of midvein, and pointed at tip. Stipules are inconspicuous (1.3–1.6 cm long), pale green to brown, sheathing, and completely free from leaf base. Floating leaves are stalked (stalks 3–5.5 cm long, continuous in color to blade junction), blades yellow green to dark green, elliptic to egg-shaped (3.5–4 cm long, 16–20 mm wide), 11–13 veins, and rounded at base. Flower clusters are dense, cylindrical (1.5–3.5 cm long) and emersed above water (flowering stem 3.2–7.7 cm long, erect, axillary and terminal). Fruit greenish brown, 1.9–2.3 mm wide, with a keel on back and on sides, and an erect beak (0.3–0.5 mm). Turions are not produced.

HABITAT & DISTRIBUTION

Potamogeton gramineus is a native northern species of shallow lakes, ponds, and rivers. It is widespread in northern New England.

Notes: Leaf shape, size, and arrangement can be highly variable in relation to water depth. Similar to Alpine pondweed (*P. alpinus*). Alpine pondweed differs by having red-tinged, longer (to 18 cm), submersed leaves, unbranched stems, and larger, plump fruits. Grassy pondweed is known to naturally hybridize with *P. illinoensis* (in VT), *P. natans* (NH), *P. nodosus* (ME), *P. oakesianus* (NH), and *P. perfoliatus* (ME, NH, VT).

136

HILL'S PONDWEED
Potamogeton hillii Morong

Hill's pondweed is a delicate, submersed, rooted, perennial, with very slender, much-branched, slightly compressed stems (30–60 cm long). Leaves are all submersed, alternate, linear (2–6 cm long, 0.6–2.5 mm wide), pale green to olive-green, 3(–5) veins, lacunae present in 1–2 rows on each side of midvein, and sharp bristle-tipped (or with an abrupt, short, projected tip). Glands usually absent at nodes, brown to green (0.1–0.3 mm wide) if present. Stipules are inconspicuous, 7–16 mm long, white to light brown, wrapped around the stem with overlapping margins, and completely free from leaf base. Flower clusters are nearly spherical (4–7 mm long), emersed above water, at end of axillary and terminal, short (6–14 mm long), recurved stems. Fruit brownish, 2.3–4 mm wide, 3-keeled, and erect beak (0.3–0.7 mm). Turions are uncommon, slender (2.8–3 cm long, 1.5–3 mm wide), soft, with leaves 2-ranked.

HABITAT & DISTRIBUTION
Potamogeton hillii is a native species of shallow, cold, mostly alkaline, lakes and streams. It is uncommon in VT.

Notes: This is the only narrow-leaved pondweed possessing leaves with <5 veins, bristle-like tips, and no nodal glands. It is similar to Leafy pondweed (*P. foliosus*).

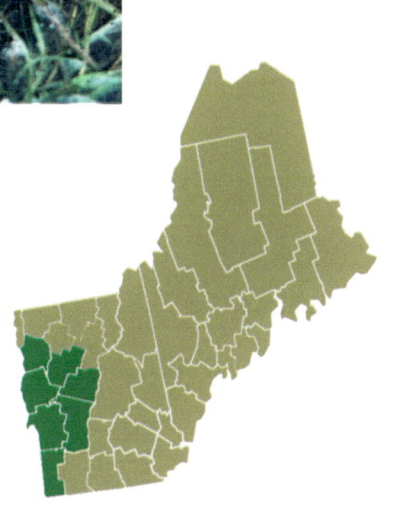

ILLINOIS PONDWEED — *Potamogeton illinoensis* Morong

Illinois pondweed is a submersed, rooted, perennial, with round, few-branched, stems (28–120 cm long), and creeping rhizomes. Leaves of two types: all submersed or both submersed and floating. Submersed leaves are alternate, stalkless or with short stalks (0.5–4 cm long), lax, flat to undulating, lance-shaped to elliptic (5–20 cm long, 20–45 mm wide), red-brown to light green, 7–15 veins, with 2–5 rows of lacunae on each side of a prominent mid-vein, and abruptly tapered to a short (to 4 mm), stiff point at tip. Stipules are conspicuous (1–8 cm long), light brown to red-brown, sheathing, and completely free from leaf base. Floating leaves are stalked (stalks 2–9 cm long), blades light green, elliptic to oblong (4–19 cm long, 2.0–6.5 cm wide), 13–29 veins, and tapering at base. Flower clusters are dense, cylindrical (2.5–7 cm long) and emersed above water (flowering stem 4–30 cm long, erect, axillary or terminal). Fruit gray green to olive-green, 2.5–3.6 mm wide, with well-developed keels and an erect to slightly curved beak (0.5–0.8 mm). Lateral keels without teeth. Turions are not produced.

HABITAT & DISTRIBUTION

Potamogeton illinoensis is a stout, native species of shallow to deep, mostly alkaline, lakes, ponds, rivers, and streams. It is common in VT.

Notes: Floating leaves are often not produced, especially in deeper water. It is similar to Long-leaf pondweed (*P. nodosus*), but with stalkless or shorter-stalked, wider, submersed leaves and green fruits. When submersed blades are stalkless, plants resemble Grassy pondweed (*P. gramineus*) but have more numerous veins and larger fruits. It is known to hybridize with Grassy pondweed, Long-leaf pondweed, and Big-leaved pondweed (*P. amplifolius*) in VT.

139

FLOATING PONDWEED *Potamogeton natans* L.

Floating pondweed is a submersed, rooted, perennial, with round, mostly unbranched, slender stems (30–90 cm long) and creeping rhizomes. Leaves both submersed and floating. Submersed leaves are alternate, stiff, linear (9–20 cm long, 0.7–2.5 mm wide), phyllodial (appearing as a bladeless leaf stalk), light to dark green, 3–5 obscure veins, lacking lacunae bands, and blunt at tip. Stipules are conspicuous (4–10 cm long), whitish, sheathing, and completely free from leaf base. Floating leaves are long-stalked (stalks 5–29 cm long, with a distinctly bent and lighter area at blade junction), blades light green, oval-elliptic to egg-shaped (5–10 cm long, 1.5–6 cm wide), 17–35 veins, usually rounded or heart-shaped at base. Flower clusters are dense, cylindrical (2.5–5 cm long) and emersed above water (flowering stem 4.5–9.5 cm long, erect, and terminal). Fruit greenish brown, 3.5–5 mm wide, wrinkled on sides, with an inconspicuous keel on back and an erect to curved beak (0.4–0.8 mm). Turions are not produced.

HABITAT & DISTRIBUTION

Potamogeton natans is a native species of acidic to alkaline lakes, ponds, rivers, and streams, including tidally influenced waters. It is widespread and common.

Notes: This plant is easily identified by its phyllodial submersed leaves (when persisting), a band of paler green tissue on end of leaf stalk where it meets a floating blade, and usually lobed base of floating blades. In non-flowing waters, the apex of leafstalks on floating leaves are noticeably bent at a 90° angle where it meets the blade. The floating leaf blade shape and degree of lobing at base is influenced by water currents. It is similar to Oakes' pondweed (*P. oakesianus*). It is known to hybridize with Grassy pondweed (*P. gramineus*) in NH.

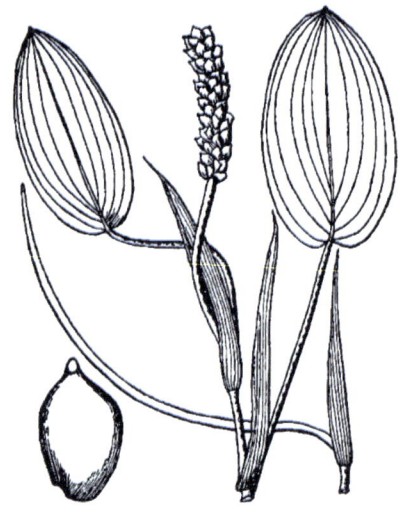

LONG-LEAF PONDWEED *Potamogeton nodosus* Poir.

Long-leaf pondweed is a submersed, rooted, perennial, with round, few-branched, stems (to 100 cm long), and whitish rhizomes. Leaves of two types: all submersed or both submersed and floating. Submersed leaves are alternate, with long stalks (5–15 cm long), lax, flat, narrowly lance-elliptic (9–20 cm long, 10–35 mm wide), light to dark green, 7–15 veins, with 2–5 conspicuous rows of lacunae on each side of a prominent mid-vein, and pointed at tip. Stipules are conspicuous (3–9 cm long), light brown, sheathing, and completely free from leaf base. Floating leaves are stalked (stalks 3.5–26 cm long), blades light green, elliptic (3–11 cm long, 1.5–4.5 cm wide), 11–21 veins, and tapering to rounded at base. Flower clusters are dense, cylindrical (2–7 cm long) and emersed above water (flowering stem stout, 3–15 cm long, erect, terminal). Fruit reddish brown, 2.7–4.3 mm wide, with well-developed keels and an erect beak. Lateral keels with blunt to sharp teeth. Turions are not produced.

HABITAT & DISTRIBUTION

Potamogeton nodosus is a native species of shallow to deep lakes, rivers, and streams. It is frequent, but rare and state-listed as *threatened* in NH. It occupies mostly low alkaline, fast-flowing waters in eastern areas and highly alkaline ponds and sluggish streams in western areas.

Notes: Plants are easily distinguished by the large, long-stalked submersed leaves and long-stalked, elliptic floating leaves, each tapering at both ends. It is similar to Illinois pondweed (*P. illinoensis*), but with longer-stalked, narrower, submersed leaves and reddish fruits. Submersed leaves often senesce before fruiting, so plants then can resemble Floating pondweed (*P. natans*). This species is known to hybridize with Grassy pondweed (*P. gramineus*) in ME and Illinois pondweed (*P. illinoensis*) in VT.

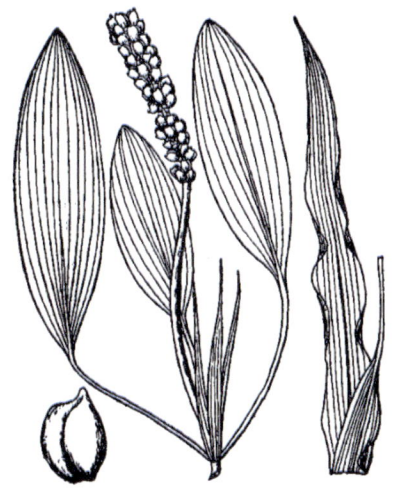

142

OAKES' PONDWEED
Potamogeton oakesianus J.W. Robbins

Oakes' pondweed is a submersed, rooted, perennial, with round, much-branched, red-spotted, slender stems (7–75 cm long) and creeping rhizomes. Leaves both submersed and floating. Submersed leaves are alternate, delicate and lax, linear (5–16 cm long, 0.25–1 mm wide), phyllodial (appearing as a bladeless leaf stalk), pale green, 3 veins, lacking lacunae bands, and pointed at tip. Stipules are conspicuous (2.5–4 cm long), whitish, sheathing, and completely free from leaf base. Floating leaves are stalked (stalks 3.2–7.5 cm long, continuous in color to blade junction), blades light to dark green, lance- to egg-shaped (2.5–6 cm long, 1–2 cm wide), 7–23 veins, and rounded or tapering at base. Flower clusters are dense, cylindrical (1–3.5 cm long) and emersed above water (flowering stem 2.5–8 cm long, erect, and terminal). Fruit greenish brown, 2–3.5 mm wide, not wrinkled on sides, with a prominent keel on back and keels on sides, and an erect beak (0.4–0.8 mm). Turions are not produced.

HABITAT & DISTRIBUTION

Potamogeton oakesianus is a native species of shallow, acidic, bogs, lakes, ponds and occasionally streams. It is widespread and common in most of northern New England.

Notes: As our only other pondweed with submersed, phyllodial leaves, it is similar to Floating pondweed (*P. natans*), but these leaves in Oakes' pondweed are much narrower (delicate and soon deteriorating too) as are the floating blades, which are not lobed at base. This species, originally described to science from material collected in Uxbridge, MA, is named after famous New England botanist William Oakes (d. 1848). It is known to hybridize with Grassy pondweed (*P. gramineus*) in ME

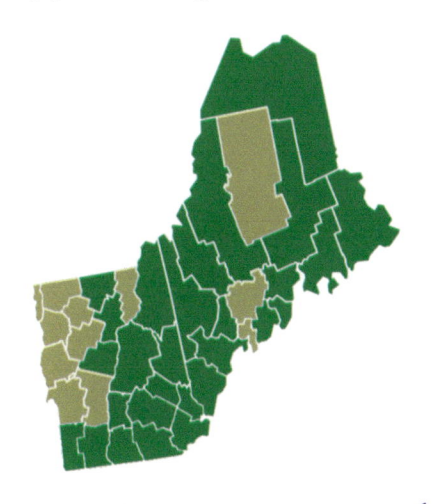

BLUNT-LEAVED PONDWEED   *Potamogeton obtusifolius* Mert. & Koch

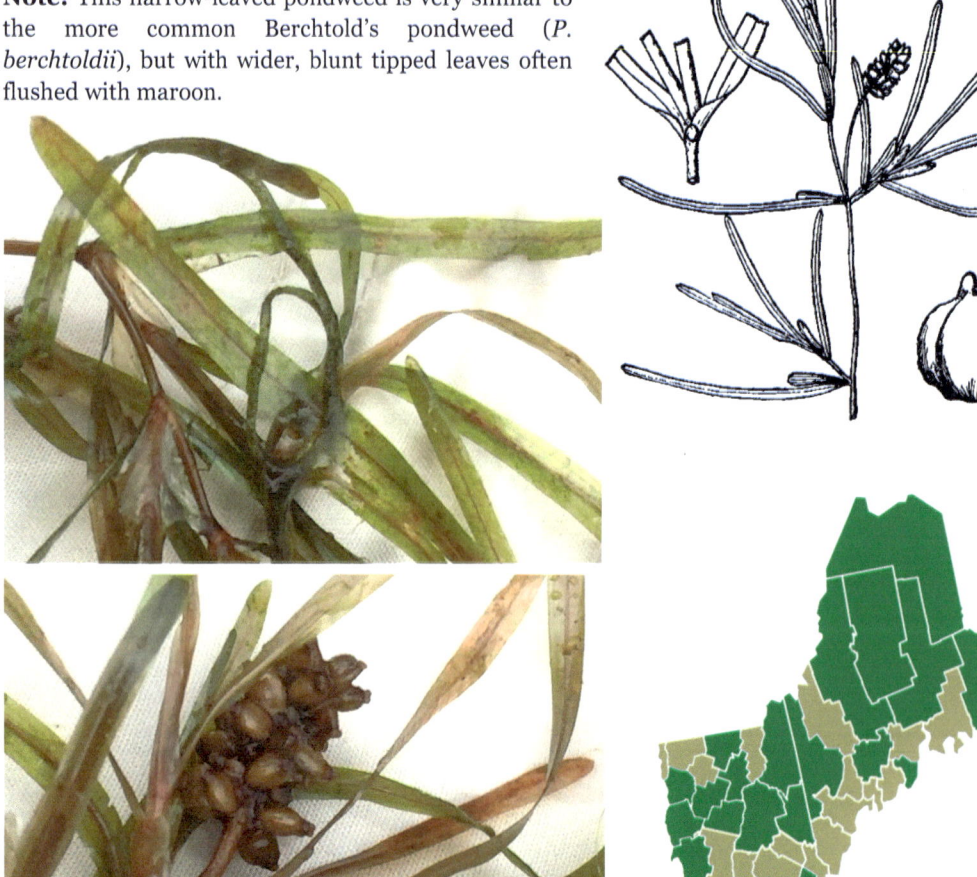

Blunt-leaved pondweed is a delicate, submersed, rooted, perennial, with slender, slightly compressed stems (35–90 cm long). Leaves are all submersed, alternate, flaccid, linear (3–8.2 cm long, 1–3.5 mm wide), green to reddish, 3 veins, lacunae present in 1–3 rows on each side of vein, and bluntly pointed or rounded at tip. Paired, circular, raised glands (0.3–1 mm wide), are yellowish green to gold, are common at nodes. Stipules are inconspicuous, 6–18 mm long, white, delicate, rarely shredding at tip, sheathing the stem, and completely free from leaf base. Flower clusters are cylindrical (8–13 mm long), emersed above water, at end of rounded stems (0.8–2 cm long). Fruit olive green to brown, 2.5–3.5 mm wide, sometimes keeled, and with an erect beak (0.8–1 mm). Turions are abundant, slender (3.5–7.8 cm long, 2.3–5.1 mm wide), soft, and leaves two-ranked, flattened, with inner and outer leaves in same plane.

HABITAT & DISTRIBUTION

Potamogeton obtusifolius is a native species of shallow, moderately calcareous waters of lakes and slow streams. It is infrequent, rare and state-listed as *endangered* in NH.

Note: This narrow-leaved pondweed is very similar to the more common Berchtold's pondweed (*P. berchtoldii*), but with wider, blunt tipped leaves often flushed with maroon.

CLASPING-LEAVED PONDWEED — *Potamogeton perfoliatus* L.

Clasping-leaved pondweed is a submersed, rooted, perennial, with round stems (to 2.5 m long) and unspotted rhizomes. Leaves submersed, alternate, lax, egg-shaped to near circular (1–7.6 cm long, 0.7–4 cm wide), with minute (1-celled) teeth on margins, blade base rounded and conspicuously clasping stem, olive green, 7–15 delicate veins, lacking lacunae, and with a flat, rounded tip. Stipules are inconspicuous (3.5–6.5 cm long), delicate, disintegrating into fibers, absent on lower part of stem, light brown to green, sheathing the stem, and completely free from leaf base. Flower clusters are cylindrical (0.4–4.8 cm long) and emersed above water, on cylindric stems (1–7.3 cm long). Fruit greenish brown, 1.6–3 mm wide, without keels, and an erect beak (0.4–0.6 mm). Turions are not produced.

HABITAT & DISTRIBUTION

Potamogeton perfoliatus is a native species of lakes, bays, and rivers, including brackish, tidal rivers. It is widespread in northern New England. It is known to hybridize with *P. alpinus* (in VT), *P. berchtoldii* (ME, VT), *P. epihydrus* (ME), *P. gramineus* (ME, NH, VT), and *P. richarsonii* (ME, VT).

WHITE-STEMMED PONDWEED   *Potamogeton praelongus* Wulfen

White-stemmed pondweed is a submersed, rooted, perennial, with round, zig-zagged stems (to 2 m long) and red spotted, white rhizomes. Leaves submersed, alternate, lax, linear to lance-shaped (6–28 cm long, 1–4.6 cm wide), blade base conspicuously clasping stem (without lobes), pale green, 11–33 veins, lacking lacunae, and with a hood-like, boat-shaped tip. Stipules are conspicuous (3–8 cm long), mostly white, fibrous, shredding at tip, sheathing the stem, and completely free from leaf base. Flower clusters are cylindrical (3.4–7.5 cm long) and emersed above water, on long (9.5–53 cm) stems. Fruit greenish brown, 4–5.7 mm wide, with a prominent, sharp keel on back, and an erect beak (0.6–0.1 mm). Turions are not produced.

HABITAT & DISTRIBUTION

Potamogeton praelongus is a native northern species of, shallow to deep, neutral to alkaline, waters of lakes and rivers. It is throughout most of northern portions, but rare and state-listed as *endangered* in NH.

Notes: White-stemmed pondweed is distinctive with its large white stipules, submersed leaves, clasping zig-zagged stems, and boat-shaped leaf tips (that split when pressed flat). It is known to hybridize with *P. amplifolius* in ME and VT.

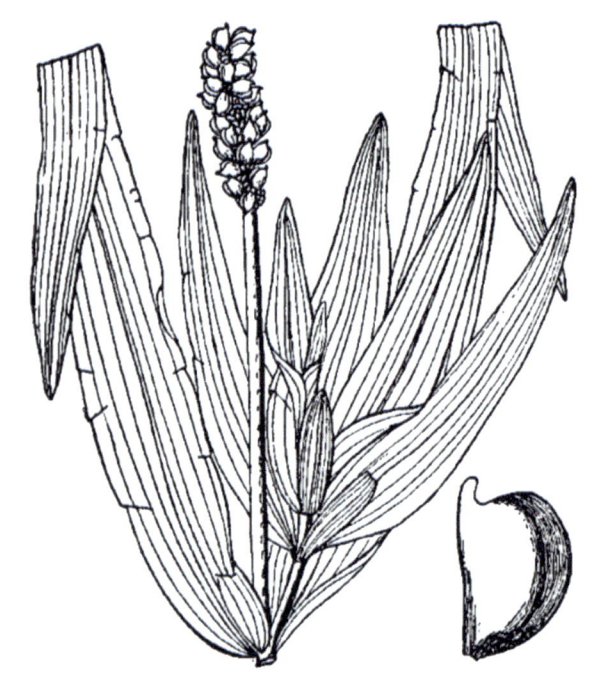

147

SPOTTED PONDWEED — *Potamogeton pulcher* Tuck.

Spotted pondweed is a submersed, rooted, perennial, with round, conspicuously black-spotted, unbranched stems (8–95 cm long) and creeping rhizomes. Leaves both submersed and floating. Submersed leaves are alternate, with narrow, black-spotted stalks (0.5–4.5 cm long), lax, slightly arching, narrowly lance-shaped (3.5–13.8 cm long, 10–25 mm wide), undulating margins, dark green, 7–19 veins, with 2–5 rows of lacunae on either side of midvein, and pointed at tip. Stipules are inconspicuous (0.7–1.2 cm long), light to dark brown, sheathing, and completely free from leaf base. Floating leaves are stalked (stalks stout, 1–16.5 cm long, black-spotted), blades light to dark green, broadly elliptic (2.5–8.5 cm long, 1.1–4.4 cm wide), 15–19 veins, and rounded at base. Flower clusters are dense, cylindrical (1.7–3.6 cm long) and emersed above water (flowering stem 3.3–9.4 cm long, erect, axillary and terminal). Fruit dark greenish brown, 3.1–4.1 mm wide, with a keel on back and sides, and an erect beak (0.5 mm). Turions are not produced.

HABITAT & DISTRIBUTION

Potamogeton pulcher is a native, eastern species of shallow, acidic lakes and rivers. It is rare in northern New England; state-listed as *threatened* in ME.

Note: The conspicuously black-spotted stems and leaf stalks are distinctive. Similar to Big-leaved pondweed (*P. amplifolius*) but with black-spotted stems, less-broad submersed blades with fewer veins, and floating blades with fewer veins. It is known to hybridize with *P. amplifolius* in NH.

SMALL PONDWEED — *Potamogeton pusillus* L.

Small pondweed is a delicate, submersed, rooted, perennial, with slender stems (18–150 cm long). Leaves are all submersed, alternate, flaccid, linear (1.4–6.5 cm long, 0.5–1.9 mm wide), pale green to olive-green, 1–5 veins, lacunae absent or present in 1–2 rows on each side of vein, and pointed at tip. Paired, circular, raised glands (to 0.5 mm wide), green (same as stem), are inconspicuous at a few nodes. Stipules are inconspicuous, 3.1–9.2 mm long, brown to green or white, delicate, not shredding at tip, with fused margins, forming a tube surrounding the stem, and completely free from leaf base. Flower clusters are cylindrical (1.5–10 mm long), usually with 2–4 interrupted whorls of flowers/fruits, emersed above water or submersed, at end of narrow, mostly terminal, stems (0.1–6.2 cm long), only 1–3 per plant. Fruit green to brown, 1.5–2.2 mm wide, not keeled, sides concave, and with an erect beak (0.1–0.6 mm). Turions are common, slender (0.9–3.2 cm long, 0.3–1.8 mm wide), soft, and leaves two-ranked, with inner leaves modified into a hardened structure.

HABITAT & DISTRIBUTION

Potamogeton pusillus is a native species of lakes, pools, and slow streams. It is uncommon in northern New England.

Notes: This narrow-leaved pondweed is distinctive with its narrow, 1–5 veined leaves, collar-like stipules, and interrupted fruit spikes (up to 3 per plant). It is similar to the more widespread Berchtold's pondweed (*P. berchtoldii*).

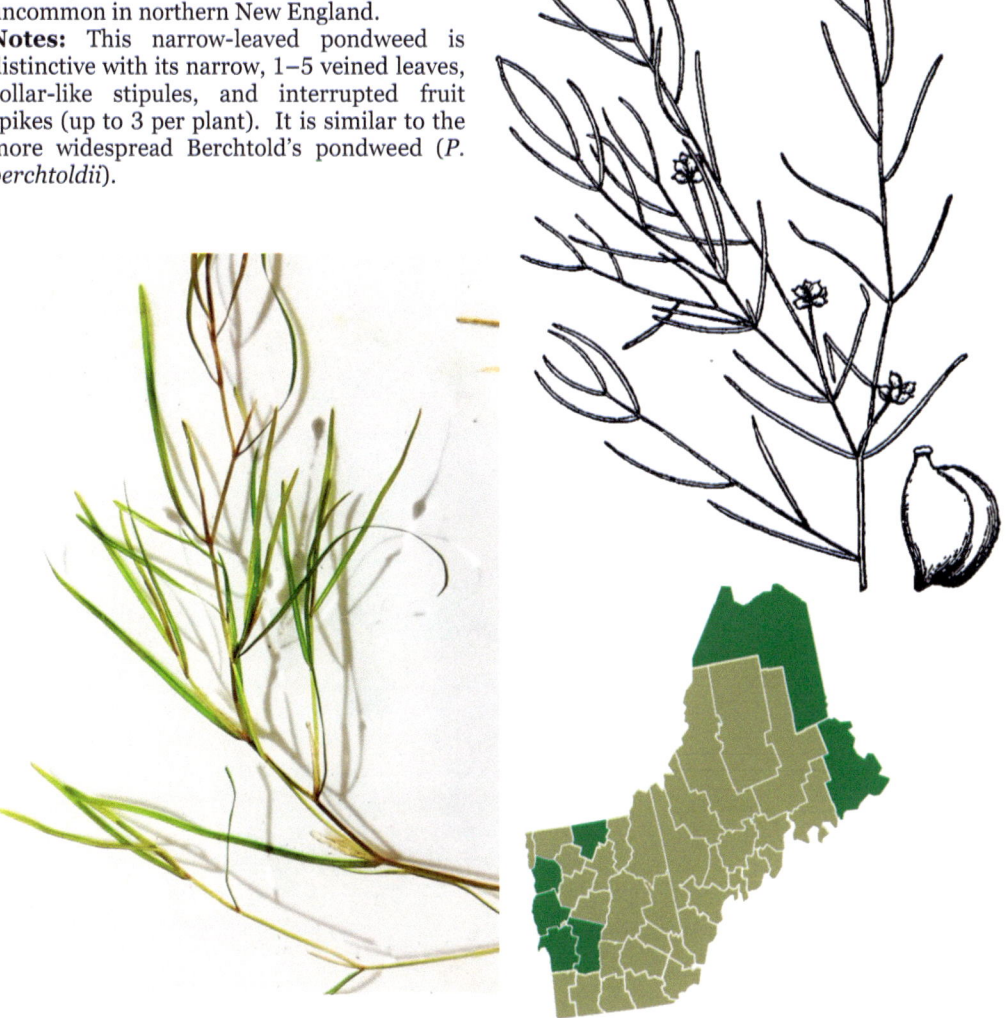

RICHARDSON'S PONDWEED *Potamogeton richardsonii* (Benn.) Rydb.

Richardson's pondweed is a submersed, rooted, perennial, with round stems (to 1 m long) and rhizomes. Leaves submersed, alternate, lax, narrowly egg-shaped to narrowly lance-shaped (1.6–13 cm long, 0.5–2.8 cm wide), with minute teeth on margins, blade base rounded and conspicuously clasping stem, olive green, 13–21 coarse veins, lacking lacunae, and with a flat, pointed tip. Stipules are conspicuous (1.2–1.7 cm long), firm, disintegrating into coarse, white fibers, sheathing the stem, and completely free from leaf base. Flower clusters are cylindrical (1.3–3.7 cm long) and emersed above water, on cylindric stems (1.5–14.8 cm long). Fruit greenish brown, 2.2–4.2 mm wide, usually without keels, and an erect beak (0.4–0.7 mm). Turions are not produced.

HABITAT & DISTRIBUTION

Potamogeton richardsonii is a native species of shallow, mostly calcareous, lakes and streams. It is frequent in northern New England, but rare and state-listed as *endangered* in NH.

Notes: It is similar to the more widespread Clasping-leaved pondweed (*P. perfoliatus*) with which is known hybridize with in ME and VT.

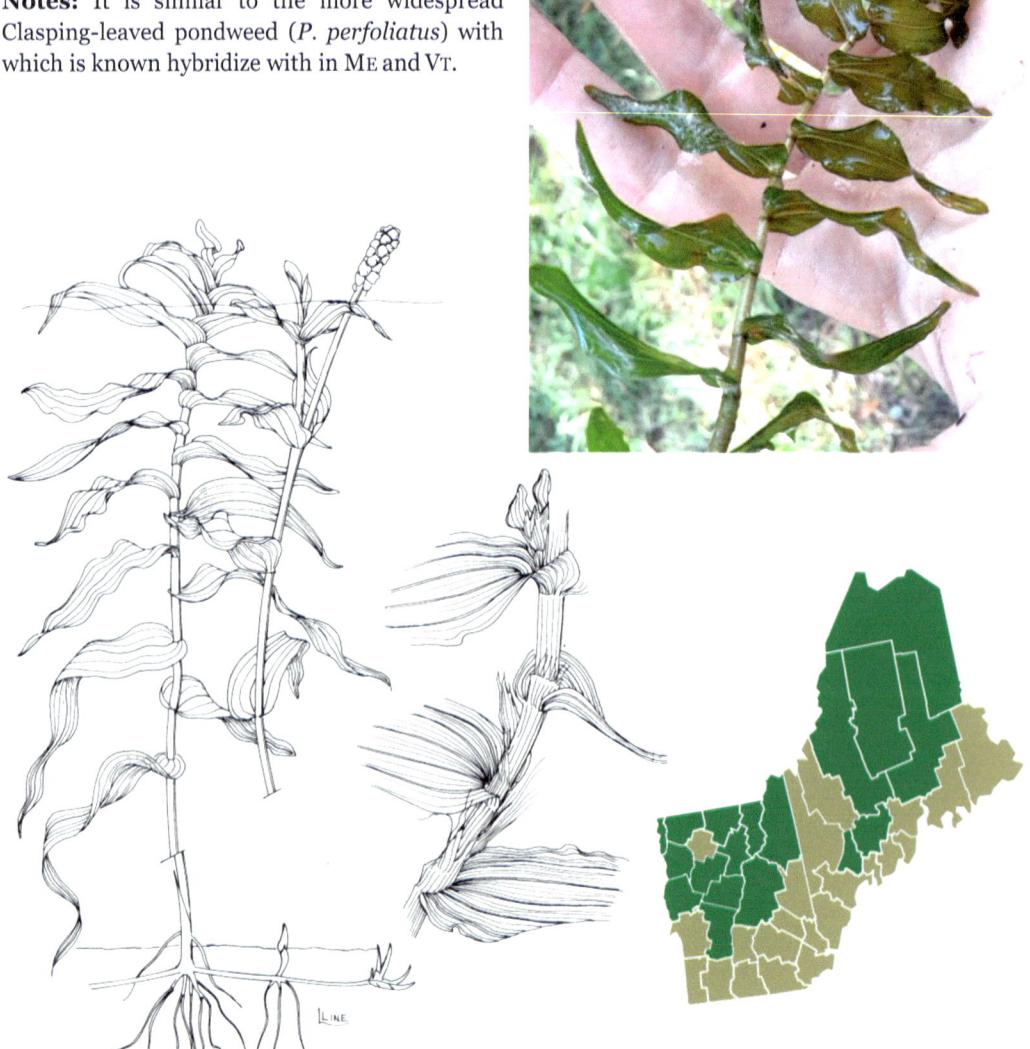

ROBBINS' PONDWEED — *Potamogeton robbinsii* Oakes

Robbins' pondweed, or Fern pondweed, is a submersed, rooted, perennial, with round, branched, slender stems (to 1 m long) and creeping rhizomes. Leaves are all submersed, alternate, strongly two-ranked (attached on opposite sides of the stem so that whole plant is flat), closely spaced, dark green to reddish, stiff, lance-shaped (20–120 mm long, 3–8 mm wide), sharply pointed, and with pair of round lobes at base (where leaf meets stipule) that protrude past the stem. Leaves are often arching, with a conspicuous midvein, and finely toothed edges. A conspicuous, translucent, greenish brown to white, stipular sheath is fused to base of leaves for ¼ of stipule length, with remaining tip of stipule free and shredding. Flowers are borne just above water in slender spikes (7–20 mm long) on branching stems. Fruits (3–5 mm long, 2–3 mm wide), rarely produced, are brown, with an erect beak (0.7–0.9 mm long) and wavy keel. Turions are not produced.

HABITAT & DISTRIBUTION

Potamogeton robbinsii is a native species of lakes and slow rivers. It is common and widespread in northern New England.

Notes: Plants are often in deep waters (5–7 m) where populations are dense (to 1000 plants per m^2). It is the only pondweed to have branched flowering stems and lobed leaf bases. It rarely flowers or sets fruit.

NORTHERN SNAIL-SEED PONDWEED *Potamogeton spirillus* Tuck.

Northern snail-seed pondweed is a submersed, rooted, perennial, with compressed, branched, slender stems (5–40 cm long) and creeping rhizomes. Leaves of two types: all submersed or both submersed and floating. Submersed leaves alternate, linear (0.7–80 cm long, 0.5–2 mm wide), red brown to light green, 1–3 veins, commonly with a broad lacunar band along each side of the midrib, and blunt-tipped. Stipules are inconspicuous, reddish brown to light green, fused to base of submersed leaves for ½ of stipule length, with remaining tip of stipule free (usually equal to or shorter than fused portion). Floating leaves are stalked (stalks 5–25 mm long), blades elliptic to lance-shaped (6–35 mm long, 2–13 mm wide), 5–15 veins, with rounded leaf tips. Flower clusters of two types: either globular (2–5 mm long) and submersed (flowering stem 0.5–3 mm long, recurved, axillary) or short-cylindrical (4–13 mm long) and emersed above water (flowering stem 4–27 mm long, erect or recurved, axillary or terminal). Fruit greenish brown, 1.5–2.5 mm wide (including keel), with blunt keel and lacking a beak. Embryo distinctly coiled in a full spiral. Turions are not produced.

HABITAT & DISTRIBUTION

Potamogeton spirillus is a native, northeastern species of lakes and rivers. It is widespread and common in northern New England.

Notes: This is one of only two species in our flora that produce two types of flowering spikes (emersed and submersed). The other species is Snail-seed pondweed (*P. bicupulatus*) which is similar to *P. spirillus*, but has thread-like submersed leaves, fused portion of stipules shorter than the free portion, and pointed floating blades.

153

STRAIGHT-LEAVED PONDWEED — *Potamogeton strictifolius* Benn.

Straight-leaved pondweed is a delicate submersed, rooted, perennial, with slender, rounded, rigid stems (27–95 cm long). Leaves are all submersed, alternate, fairly stiff, linear (1.2–6.3 cm long, 0.6–2 mm wide), green to olive-green, 3–5 veins, lacunae absent, and sharp-pointed to nearly bristle-tipped. Paired, circular, raised glands (to 0.3 mm wide), are white, green, greenish brown, or gold, commonly at nodes. Stipules are inconspicuous, 6–16 mm long, white, fibrous, shredding at tip, with fused margins forming a cylinder or tube around stem, and completely free from leaf base. Flower clusters are cylindrical (6–13 mm long), emersed above water, at end of rounded stems (1–4.5 cm long). Fruit green-brown, 1.9–2.1 mm wide, not keeled, and with an erect beak (0.5–0.8 mm). Turions are common, slender (2.5–4.8 cm long, 1–2.2 mm wide), soft, and leaves two-ranked, flattened, with inner and outer leaves in same plane.

HABITAT & DISTRIBUTION

Potamogeton strictifolius is an uncommon, native species of shallow, calcareous waters of lakes and slow streams. It is rare in northern New England; state-listed as *threatened* in ME

Note: The rigid leaves of this narrow-leaved pondweed hold their shape when removed from water. It is known to hybridize with Flat-stem pondweed (*P. zosteriformis*) in VT.

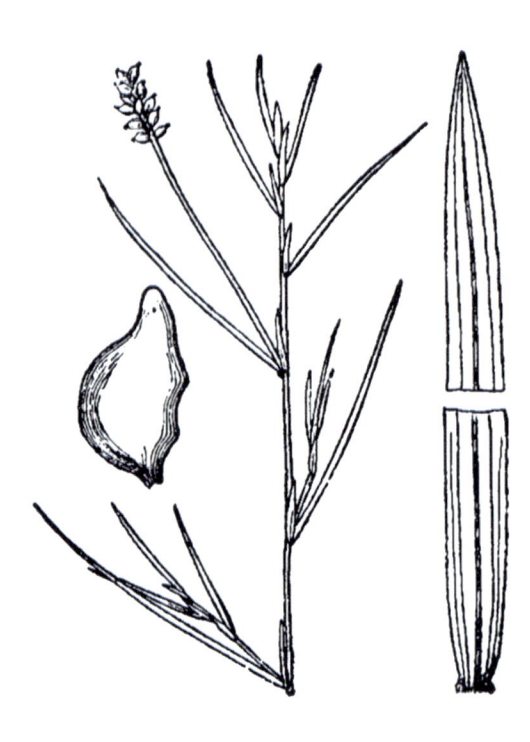

VASEY'S PONDWEED — *Potamogeton vaseyi* J.W. Robbins

Vasey's pondweed is a very delicate submersed, rooted, perennial, with round, slender stems (10–25 cm long). Leaves of two types: all submersed or both submersed and floating. Submersed leaves alternate, linear and thread-like (2–8 cm long, 0.1–1 mm wide), light green, 1(–3) vein, lacunae absent, and pointed or long-tapering at tip. Stipules are inconspicuous, green to brown, and completely free from leaf base. Floating leaves are stalked (stalks 5–25 mm long), blades elliptic to oblong, small (6–15 mm long, 3–8 mm wide), 5–9 veins, with blunt leaf tips. Flower clusters short-cylindrical (3–8 mm long) at end of stems (5–30 mm long), emersed above water, and erect (or recurved in fruit). Fruit is greenish brown, 1.5–2.5 mm long, 1–1.5 mm wide, with a smooth keel and erect beak (0.3–0.5 mm), and embryo distinctly coiled with a full spiral. Turions are common in leaf axils as slender (5–20 mm long, 0.5–1.2 mm wide), soft, two-ranked outer leaves and rolled, hardened inner leaves.

HABITAT & DISTRIBUTION

Potamogeton vaseyi is a native species of shallow, acidic to alkaline, lakes and rivers. It is extremely rare and state-listed as *special concern* (ME) and *endangered* (NH).

Notes: Floating leaves are present only when plants are flowering and then plants deteriorate by mid-August. When floating leaves are absent, this plant is difficult to distinguish from Small pondweed (*P. pusillus*) or Budding pondweed (*P. gemmiparus*) with which it often co-occurs.

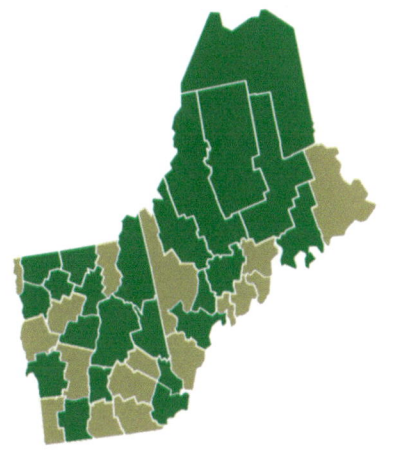

FLAT-STEM PONDWEED *Potamogeton zosteriformis* Fern.

Flat-stem pondweed is a submersed, rooted, perennial, with conspicuously flattened (appearing winged), stiff, stems (60–120 cm long, 0.6–3.2 mm wide). Leaves are all submersed, alternate, linear (10–20 cm long, 2–5 mm wide), light green, 15–35 veins, lacunae absent, and blunt to tapering at tip. Circular glands (to 0.3 mm wide), gold, are sometimes present at nodes. Stipules are conspicuous, 15–35 mm long, firm, fibrous, white, shredding at tip, sheathing the stem, and completely free from leaf base. Flower clusters are cylindrical (15–30 mm long), emersed above water, at end of stout, cylindric stems (2–5 cm long). Fruit olive green to green, 4–4.5 mm wide, sharp keeled on back, and with an erect beak (0.6–1 mm). Turions are common, flattened, large (4–7.5 cm long, 2–4.5 mm wide), firm, with leaves two-ranked, the inner and outer leaves oriented in the same plane, and the inner leaves undifferentiated

HABITAT & DISTRIBUTION

Potamogeton zosteriformis is a native species of shallow lakes, ponds, and streams. It is widespread in much of northern New England, but rare and state-listed as *endangered* in NH.

Notes: It is a distinctive pondweed with its flat, winged stems and many-veined, parallel-sided, linear leaves to 5 mm wide. Leaves hold their shape when removed from water. It is known to hybridize with Straight-leaved pondweed (*P. strictifolius*) in VT.

Proserpinaca — Haloragaceae
Mermaid-weed

The Mermaid-weeds are amphibious, rooted, perennials that can be mostly or entirely submersed. Stems, rooted from the lower nodes, grow up out of water from rhizomes. Leaves are alternate, and their form varies greatly by water depth: blades range from smaller and deeply pinnately divided (submersed) to larger and merely toothed (emergent). Flowers are small, purplish-greenish, and borne 1–3 in the emergent leaf axils. Fruit is a 3-seeded, 3-angled nutlet.

Two species occur in northern New England, both native.

KEY TO SPECIES

1. Emersed leaves (20–40 mm long) sharply-toothed, submersed leaves (5–30 mm long) deeply-divided into 7–14 pairs of segments; fruits 2.3–6 mm wide............................*P. palustris*
1. All leaves (10–30 mm long) deeply-divided, submersed leaves with 6–9 pairs of segments; fruits 2–2.8 mm wide…………………………………………………………............. *P. pectinata*

MARSH MERMAID-WEED
Proserpinaca palustris L.

Marsh mermaid-weed is a submersed plant (10–40 cm long) with flowering portions emergent. Leaves are alternate (20–40 mm long), emersed ones sharply-toothed and submersed ones deeply-divided into 7–14 pairs of narrow segments 5–30 mm long. Flowers are borne in axils of emersed leaves, each with a toothed leaf-like bracts below. Fruits are 2.3–6 mm wide and sharply angled.

HABITAT & DISTRIBUTION

Proserpinaca palustris is native to shallow ponds and slow streams. It is widespread in northern New England.

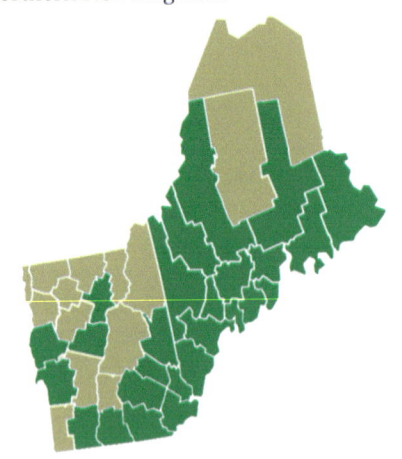

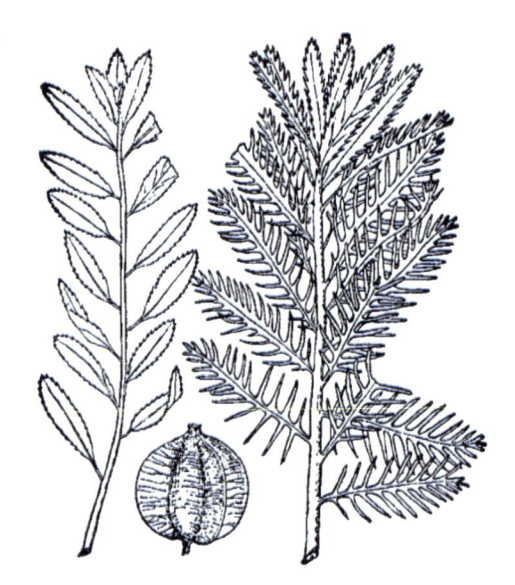

COMB-LEAVED MERMAID-WEED   *Proserpinaca pectinata* Lam.

Comb-leaved mermaid-weed is usually a submersed plant with flowering portions emergent. Leaves are alternate (10–30 mm long) and deeply divided into 6–9 pairs of narrow segments 2–7.5 mm long. Flowers are borne in axils of emersed leaves, each with a deeply divided leaf-like bract. Fruits are 2–2.8 mm wide and bluntly angled.

HABITAT & DISTRIBUTION

Proserpinaca pectinata is native to shallow ponds, bogs, and ditches. This Coastal Plain species reaches its northern limits in northern New England. It is rare and state-listed as *endangered* in both ME and NH.

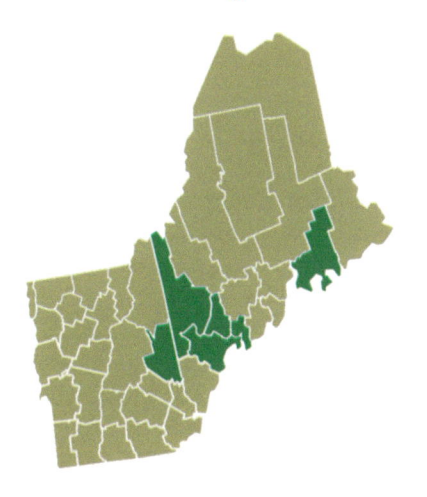

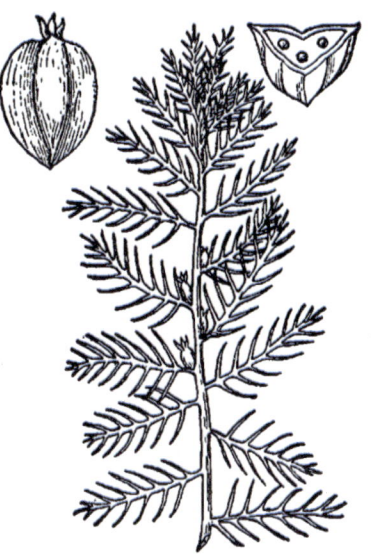

159

Ranunculus — Ranunculaceae
Crowfoot

The Crowfoots, or Buttercups, have just a few truly aquatic species with highly dissected, submersed leaves (or deeply lobed floating leaves). Leaves are obviously stalked and expanded at base sheathing the stem. Flowers are conspicuous and held above water, each with 5 colored petals, and numerous stamens and pistils. Fruits are clusters of small, seed-like achenes.

Four aquatic species occur in northern New England, all native and highly variable in terms of leaf shape depending on conditions. Additionally in the region there are several other species of Buttercups (e.g., *R. recurvatus* Poir., *R. repens* L., and *R. cymbalaria* Pursh) that occupy wetland shores. These emergent species (not detailed here) have mostly erect stems and simple, broad leaves held above water.

KEY TO SPECIES

1. Leaves all arising from base of plant (basal) on long stalks, blades floating (or emersed), flat, and deeply lobed..*R. sceleratus*
1. Leaves alternate along stems, mostly submersed and highly divided into very narrow segments..2
   2. Leaf segments flat and ribbon-like; petals yellow; fruits smooth....................... *R. flabellaris*
   2. Leaf segments thread-like, petals white (with yellow blotch at base); fruits transversely wrinkled..3
3. Leaves flaccid and collapsing when out of water, leafstalk evident above sheathing base..... ....*R. trichophyllus*
3. Leaves firmer and holding shape when out of water, leafstalk absent with leaves branching immediately above sheathing base...*R. longirostris*

R. trichophyllus

R. flabellaris

YELLOW WATER CROWFOOT — *Ranunculus flabellaris* Raf.

Yellow Water crowfoot is a submersed perennial with floating, hollow, stems rooted at lower nodes. Leaves are alternate, mostly submersed, kidney-shape to semicircular in outline (1.5–7 cm long), highly divided into flat, ribbon-like segments (1–2 mm wide), with expanded, sheathing stalks. Flowers are bisexual, emersed above water, borne 1-few on a stout stalk, with 5–6, shiny, yellow petals (7–12 mm long), and green centers. Fruit small (2.5–3.5 mm long), smooth, with a winged margin, and beak (1–2 mm long).

HABITAT & DISTRIBUTION

Ranunculus flabellaris is a native species of open, shallow (30–100 cm deep), acidic or alkaline, waters of ponds, bogs, swamps, and streams. It is uncommon in northern New England.

Note: Leaves near stem tips are sometimes emersed or floating. These leaves can be merely lobed, rather than the highly divided submersed leaves.

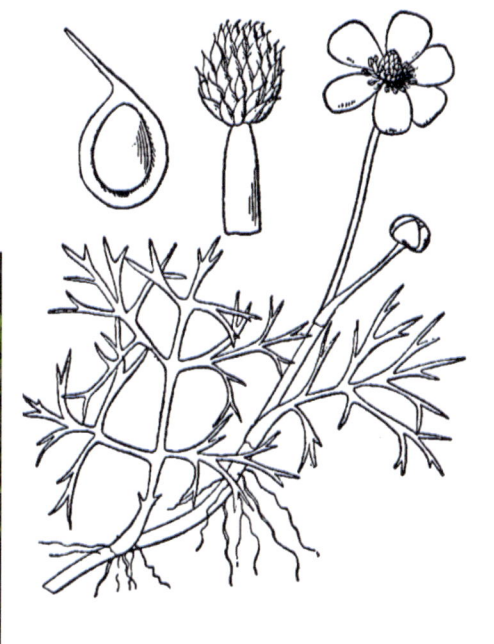

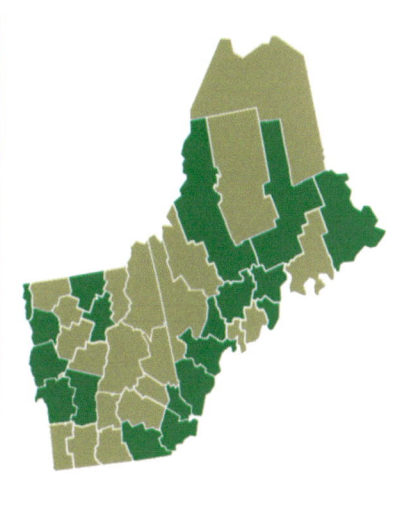

WHITE WATER CROWFOOT — *Ranunculus longirostris* Godr.

White Water crowfoot is a submersed perennial with elongate, branched stems rooted at lower nodes. Leaves are alternate, submersed, round to semicircular in outline (1.5–3 cm long), rigid, branching immediately above expanded, sheathing base, and highly divided into thread-like segments. Flowers are bisexual, emersed above water, borne singly on a long stalk from a leaf axil, with five white petals (6–8 mm long), and yellow centers. Fruit small (2 mm long), transversely wrinkled, and with persistent beak (1 mm long).

HABITAT & DISTRIBUTION

Ranunculus longirostris is a native species of cold, deep (2–5 m), ponds and slow streams. It is restricted to mostly alkaline waters of Vermont.

Notes: Due to differences of taxonomic opinion, this plant has often been combined with the Eurasian *R. aquatilis* L. in other references. It is similar to *R. trichophyllus*.

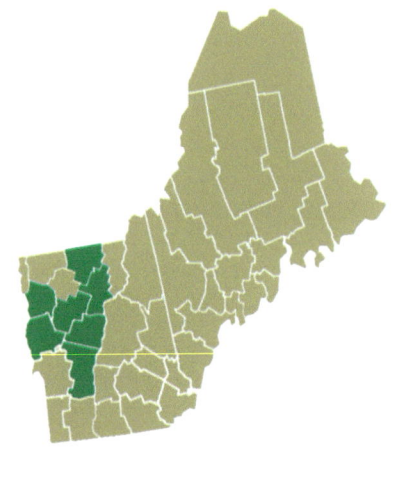

CURSED CROWFOOT   *Ranunculus sceleratus* L.

Cursed crowfoot is a submersed annual with floating hollow stems rooted at base. Leaves are all basal (arising from submersed base of plant), with long stalks, and floating, deeply lobed blades (divided into 3–5 main segments). Flowers are bisexual, emersed above water on branching clusters, with five, small (2–5 mm long), yellow petals, and green centers. Fruit small (1.5–2 mm long), transversely finely wrinkled, with a minute beak.

HABITAT & DISTRIBUTION

Ranunculus sceleratus is a presumed native of shallow, fresh to brackish waters of marshes, swamps, and tidal marshes. It is uncommon in northern New England.

Note: This species is highly variable and adaptable. Described here is the submersed form of the species. More often, plants will be erect, largely emergent, with alternate leaves supported on a hollow stem.

163

WHITE WATER CROWFOOT — *Ranunculus trichophyllus* Chaix

White Water crowfoot is a submersed perennial with elongate, branched stems rooted at lower nodes. Leaves are alternate, submersed, round to semicircular in outline (2–6 cm long), highly divided into thread-like segments (sometimes hairy), and leafstalks (5–20 mm long) with expanded, sheathing bases. Flowers are bisexual, emersed above water, borne singly on a long stalk from a leaf axil, with five white petals (3–5 mm long), and yellow centers. Fruit small (2 mm long) and transversely wrinkled.

HABITAT & DISTRIBUTION

Ranunculus trichophyllus is a native northern species of springs, ponds, and slow streams. It is widespread in northern New England.

Notes: Due to differences of taxonomic opinion, this plant has often been combined with the Eurasian *R. aquatilis* L. in other references. It is similar to *R. longirostris*.

Rorippa — Brassicaceae
Yellow Cress

LAKE YELLOW CRESS   *Rorippa aquatica* (Eat.) Palmer & Steyerm.

Lake yellow cress, or Lake cress, is a rooted, perennial with mostly submersed, flaccid stems or with upper ends emergent. Leaves are highly variable, alternate, and of two types (2–6 cm long, 7–15 mm wide): emergent with oval to oblong, simple blades, and submersed leaves highly dissected into fine thread-like divisions. Flowers are bisexual, each with four, white petals (6–10 mm long) and borne on elongated clusters above water. Fruit, if made, are oblong (4–7 mm long, 3 mm wide).

HABITAT & DISTRIBUTION

Rorippa aquatica is a native species of streams, lakes, ponds. It is rare and state-listed as *threatened* in VT. It occupies clean, calcareous, slow-moving waters that experience natural water level fluctuations. Plants exhibit high levels of vegetative reproduction whereby easily detached submersed leaves (or stems fragments) float and develop into plantlets.

Note: The introduced and more widespread Watercress (*Nasturtium officionale* Ait. f.), which also occupies shallow, clear, flowing waters, is somewhat similar. Watercress however has pinnately compound leaves and longer (10–18 mm) fruits.

Ruppia — Ruppiaceae
Wigeon-grass

WIGEON-GRASS — *Ruppia maritima* L.

Wigeon-grass, or Beaked ditch-grass, is a delicate, rooted, entirely submersed, marine perennial with rhizomes and thread-like stems (<1 mm wide). Leaves are alternate, thread-like (6–10 cm long, <0.5 mm wide), green, blunt-tipped, and fused to translucent stipules which sheath the stems at nodes. Flowers are bisexual, underwater, borne in pairs within a leaf sheath when young. Fruits ovoid (2 mm long), fleshy, short-beaked, each borne on a long stalk (1–2 cm long), in clusters of four at end of a stalk (<25 mm long) which can sometimes coil.

HABITAT & DISTRIBUTION
Ruppia maritima is a native, salt-tolerant plant of ponds, streams, ditches, and tidal pools. Common in shallow, brackish and saline waters of coastal ME and NH.

Notes: Can be confused with Horned-pondweed (*Zannichellia palustris*).

Sagittaria — Alismataceae
Arrowhead

The Arrowheads are rooted, submersed, floating-leaved, or emersed perennials, often with two leaf types on same plant. Leaves arise from a basal rosette and are highly variable in shape, influenced by water depth. Roots are septate and fibrous. Flowers are showy, insect-pollinated, with three white petals, and mostly unisexual (grouped on same plant), clustered in bracted whorls of three around a usually upright stem. The upper whorls are male flowers and lower female/fruiting (or bisexual). Flattened, winged, seed-like fruits (achenes) are crowded into a spherical cluster with persistent sepals at the base.

Six truly aquatic species, all native, occur in northern New England. The widespread Common arrowhead (*S. latifolia* Willd.) occupies shallow pond margins and wetlands in the region too. This emergent species (excluded here), which can be very common, has firm, erect, foliage with obvious blades, commonly lobed at their bases.

KEY TO SPECIES

1. Sepals closely appressed in fruit, enclosing the fruit head; stalk of fruit head recurved...........2
1. Sepals spreading or recurved in fruit, not enclosing the fruit head; stalk of fruit head ascending to spreading..3
    2. Leaves flattened, long and flexuous (>30 cm long, 0.1–1.5 cm wide), either phyllodial and, or floating with a linear-ovate blade; flowering stem usually floating with 4-10 whorls of, lower flowers female..*S. filiformis*
    2. Leaves thick and spongy (≤17 cm long, 0.5–1 cm wide), phyllodial, and submersed, rarely emersed; flowering stem emersed with 1-2 whorls of flowers, lower flowers bisexual.......... .... *S. montevidensis*
3. Stamens with filaments covered in short hairs; leaf blades (if present) without lobes..........4
3. Stamens with hairless filaments; leaf blades with basal lobes..................................*S. cuneata*
    4. Leaves all phyllodial, essentially round in cross-section, generally tapering from near the base to tip... *S. teres*
    4. Leaves either phyllodial and flattened with parallel margins that taper somewhat abruptly to the apex, or with a flattened, expanded blade at tip..5
5. Lowest whorl of female flowers borne on 1–3 cm long stalks; flowering stem straight, without a distinct bend; fruits 1.5–2 mm long... *S. graminea*
5. Lowest whorl of female flowers stalkless or short-stalked (≤0.5 cm long); flowering stem often with a conspicuous bend at the lowest whorl of flowers; fruits 2.5–4 mm long.......... .... *S. rigida*

168

NORTHERN ARROWHEAD — *Sagittaria cuneata* Sheldon

Northern arrowhead is a variable species with submerged, flattened (2.5–45 cm long, 1.5–11 cm wide), phyllodial leaves and floating (or emersed) leaves. Floating leaves have expanded blades (75–90 mm long, 35–40 mm wide), usually with basal lobes (shorter than rest of blade), and 3-sided petioles. Flowers (to 25 mm wide) are emersed, with short petals (7–10 mm long), and recurved sepals that do not enclose the flower or fruit head. Stamens have linear, hairless filaments. Fruit (1.8–2.6 mm, 1.3–2.5 wide) is keeled with a diagnostic short (0.1–0.5 mm long), erect beak at the apex.

HABITAT & DISTRIBUTION

Sagittaria cuneata is a native, northern, perennial species of cold (often calcareous), pools, lakes, and sluggish streams. It is rare and state-listed as *endangered* in NH. Often occurs in open, deep waters (>1 m), and when flowing, plants take on a growth form with mostly long, broad, ribbon-like submersed leaves and relatively few floating leaves.

Notes: Our only arrowhead with truly floating leaves. Similar to the widespread Common arrowhead (*S. latifolia*), but the submersed leaves, floating blades, shorter petals (≤10 mm), and ascending, short (≤5 mm) fruit beak of *S. cuneata* distinguish it. When plants are sterile, its ribbon-like leaves may be difficult to distinguish from *Vallisneria* and *Sparganium*.

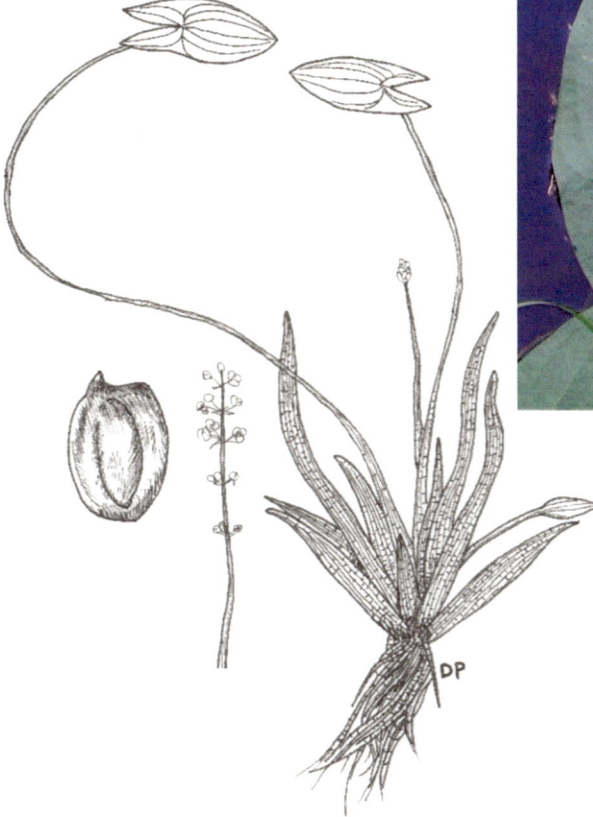

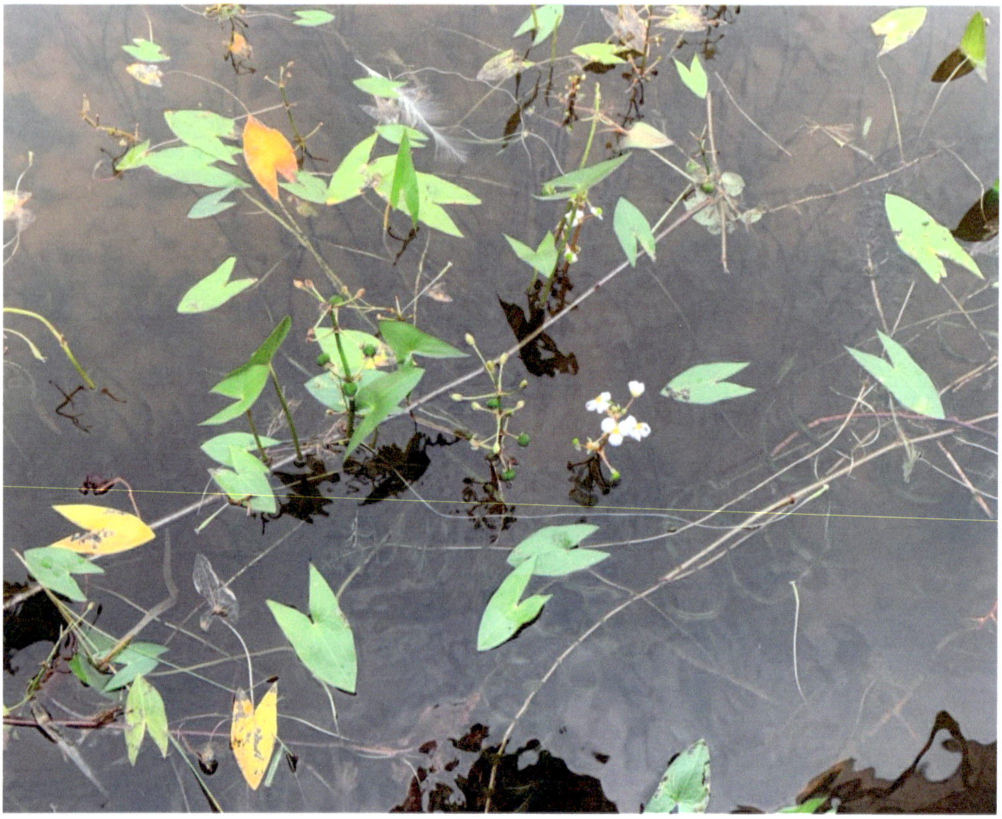

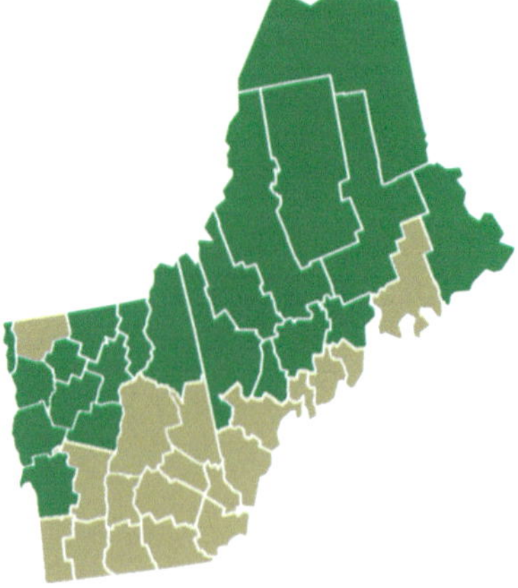

NARROW-LEAVED ARROWHEAD *Sagittaria filiformis* JG Sm.

Narrow-leaved arrowhead is a submerged perennial riverine plant. Leaves are submersed, ribbon-like, and long (30–250 cm long, 0.1–1.5 cm wide), or sometimes floating at ends with a dilated linear-ovate blade (to 0.5 cm wide). Flowers (to 30 mm wide) borne on a long (10–200 cm), usually floating, stem with 4–10 whorls of flowers. Sepals are erect and closely appressed in female flowers, enclosing the flower (or fruit head). Stamens have dilated, hairless filaments. Stalk of fruit head recurved, 1.5–4.5 cm long. Fruits have an erect (1 mm) lateral beak.

HABITAT & DISTRIBUTION

Sagittaria filiformis is a native Coastal Plain species reaching its northern limits in our area. It occupies shallow to deep reaches of slow to swift flowing waters, including tidal freshwaters. Rare and state-listed in ME (*special concern*) and NH (*endangered*).

Notes: This species is not known to produce fruit in our area.

GRASS-LEAVED ARROWHEAD — *Sagittaria graminea* Michx.

Grass-leaved arrowhead is a submerged (or emergent) rhizomatous perennial. Leaves are submersed, flattened, and phyllodial (6.5–35 cm long, 0.5–1 cm wide) with parallel margins that taper somewhat abruptly to the tip, or can have emersed leaves with a flattened, expanded blade (linear to linear-oval in shape, 2.5–17.5 cm long, 0.2–4 cm wide) lacking lobes. Flowers (to 23 mm wide) borne on an emergent, straight (without a distinct bend), stem with 1–12 whorls of flowers. Sepals are spreading or recurved in fruit, not enclosing the fruit head. Lowest whorl of female flowers borne on stalks (1–3 cm long). Stamens have dilated filaments covered in short hairs. Stalk of fruit head ascending to spreading. Fruits have an erect (0.2 mm) lateral beak.

HABITAT & DISTRIBUTION

Sagittaria graminea is a native species of ponds, rivers, and freshwater tidal marshes. Widespread in northern New England. In our region, it often occurs as submerged rosettes without flowers/fruit.

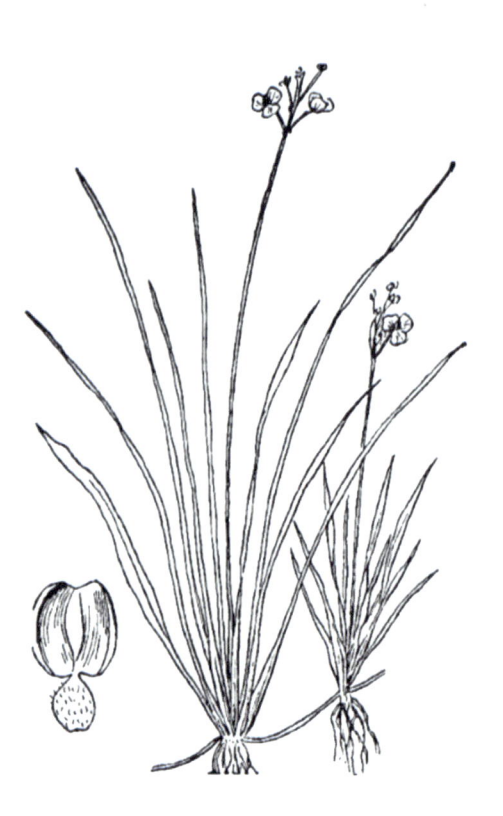

173

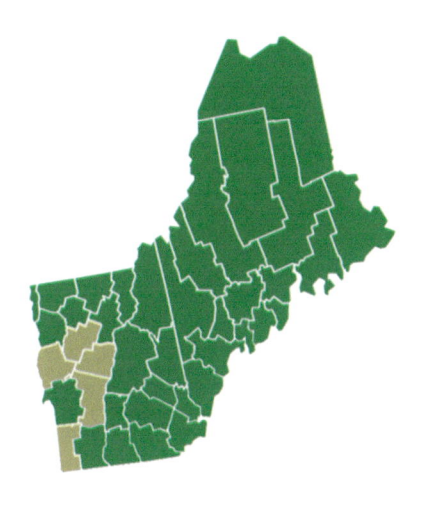

SPONGY-LEAVED ARROWHEAD — *Sagittaria montevidensis* Cham. & Schlecht.

Spongy-leaved arrowhead, or Estuary arrowhead, is a submersed, diminutive, annual plant. Leaves are submersed, thick, spongy, and strap-like (4–17 cm long, 0.5–1 cm wide). Leaves sometimes have expanded, spongy, lance-shaped blades. Flowers (ca. 20 mm wide) borne on an emersed stem, shorter than the leaves, singly or with 1–2 whorls of flowers. The lowest flowers are bisexual (with ring of sterile stamens). Sepals are erect and closely appressed in female flowers, enclosing the flower (or fruit head). Stamens have linear, hairless filaments. Stalk of fruit head thickened and recurved, 0.5–4.2 cm long. Fruits have a horizontal lateral beak (0.4–0.8 mm).

HABITAT & DISTRIBUTION

Sagittaria montevidensis is a native, coastal plant found on mudflats and shores of freshwater tidal rivers, typically exposed on mud flats at low tide. It is extremely rare and state-listed as *special concern* (ME) and *endangered* (NH).

Note: Our plants represent ssp. *spongiosa* (Engelm.) Bogin.

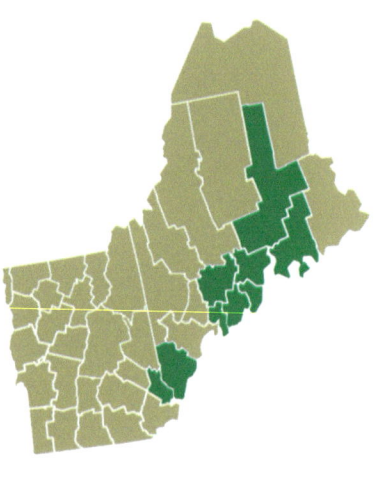

SESSILE-FRUITED ARROWHEAD — *Sagittaria rigida* Pursh

Sessile-fruited arrowhead is a submerged (or emergent) perennial with slender stolons. Leaves are submersed, stiff, flattened, and phyllodial (30–70 cm long, 0.5–1 cm wide) with parallel margins that taper somewhat abruptly to the tip, or can be emersed with an expanded blade (linear to oval in shape, 5–15 cm long, 0.6–12 cm wide) commonly lacking lobes. Flowers (to 30 mm wide) borne on an emergent stem, conspicuously bent at the lowest whorl of flowers, with 2–8 whorls of flowers. Sepals are recurved and not enclosing the flower or fruit head. Lowest whorl of female flowers stalkless or borne on very short stalks (≤0.5 cm long). Stamens have dilated filaments covered in short hairs. Fruits have a recurved, lateral beak (0.8–1.4 mm).

HABITAT & DISTRIBUTION

Sagittaria rigida is a native species with highly variable leaves (influenced by water depth) which occurs in shallow to deep, fresh or brackish, waters of rivers, ponds, and pools. Plants are uncommon; rare and state-listed as *special concern* (ME) and *endangered* (NH).

Note: When plants occur only as submersed, sterile rosettes it is impossible to distinguish from Grass-leaved arrowhead (*S. graminea*).

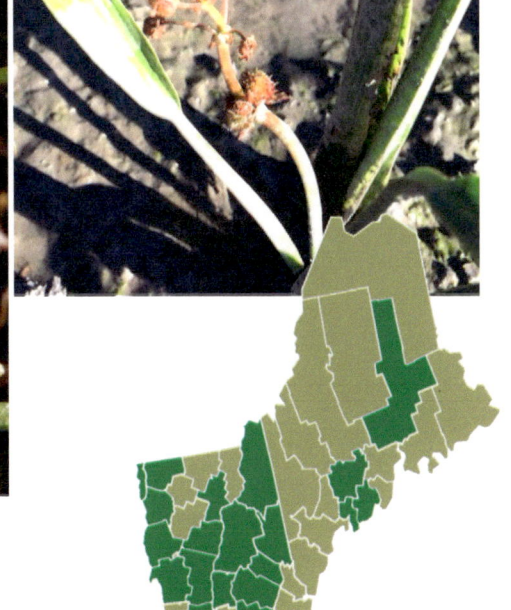

QUILL-LEAVED ARROWHEAD — *Sagittaria teres* S. Wats.

Quill-leaved arrowhead is a submerged (or emergent) perennial with slender stolons. Leaves are submersed, rigid, linear (3.5–18.5 cm long, 0.15–0.4 cm wide), and essentially round in cross-section, thickest at base generally tapering to the tip, or can have longer leaves (to 60 cm) with emersed tips. Flowers (to 15 mm wide) borne on an emergent, straight (without a distinct bend), stem with 1–4 whorls of flowers. Sepals are spreading or recurved in fruit, not enclosing the fruit head; Lowest whorl of female flowers borne on stalks (1–3 cm long). Stamens have dilated filaments covered in short hairs. Stalk of fruit head ascending to spreading. Fruits have an erect (0.3–0.4 mm) to horizontal beak.

HABITAT & DISTRIBUTION

Sagittaria teres is a native Coastal Plain species of mostly sandy, acidic ponds. It is extremely rare and state-listed as *endangered* in NH. Plants can be locally abundant and often inconspicuous as non-flowering, submersed rosettes.

Notes: Most similar to Grass-leaved arrowhead (*S. graminea*), especially when occurring as submersed basal rosettes, but leaves are always round in cross-section and never form blades.

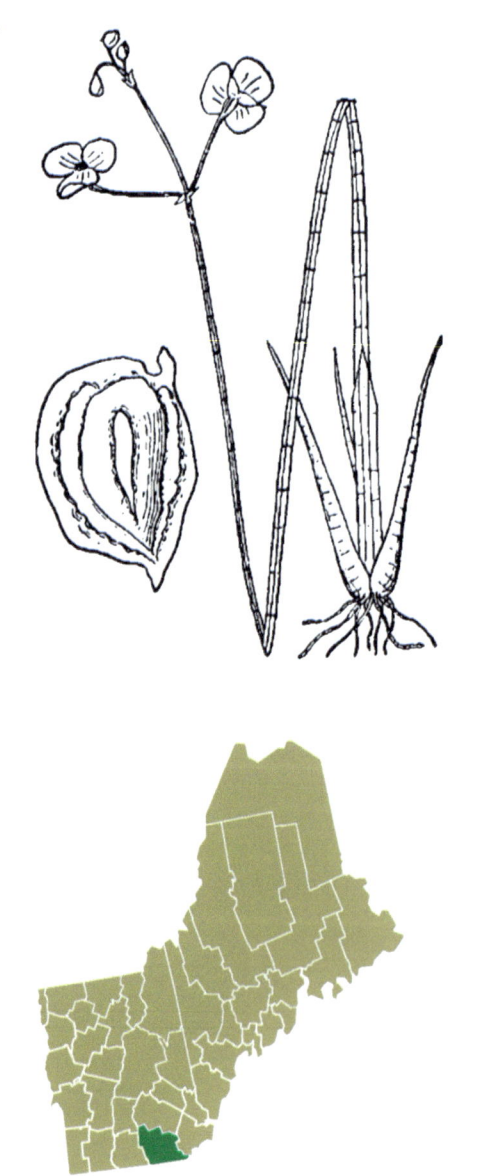

Schoenoplectus — Cyperaceae
Bulrush

WATER BULRUSH — *Schoenoplectus subterminalis* (Torr.) Soják.

Water bulrush is a rooted, grass-like, perennial, with submerged, flaccid, thread-like leaves clustered along slender (1 mm diameter), soft rhizomes and erect flowering stems. Leaves are alternate, submerged (or tips floating), 3–20, thin (to 1 mm wide), flexuous, with sheathing bases. Flowers are minute, borne in a small (4–5 mm long), straw-colored cluster at tip of a slender, upright stem (20–150 cm long), extending above water surface. A single, leaf-like (2–20 mm long) bract is attached below flower cluster, appearing like a continuation of the stem. Fruit is a seed-like (2.5 mm long), 3-sided achene.

HABITAT & DISTRIBUTION

Schoenoplectus subterminalis is a native species of lakes, bogs, and slow-moving streams. It is widespread in northern New England.

Notes: This plant typically forms dense colonies and can colonize water to 1 m deep. Sometimes leaves can be erect and C-shaped in cross-section. Similar to Robbins' spikerush (*Eleocharis robbinsii*).

Sparganium — Typhaceae
Bur-reed

The Bur-reeds are rooted, perennials with buried rhizomes and long (10–50+ cm), floating stems. Leaves are submersed or partially floating, long (0.5–1+ m), linear (2–10 mm wide), flaccid, and generally flattened. Flowers are small, wind-pollinated, and unisexual (both on same plant), grouped into dense, spherical clusters (heads) held above water. Female flowers occupy the lower positions on the flowering stem and male flowers the uppermost positions. Fruits are a spherical aggregate of small, hard, beaked drupes.

   Three aquatic species, all native, occur in northern New England. Mature fruits are needed for correct identification. Several other Bur-reed species (e.g., *S. americanum* Nutt., *S. androcladum* (Engelm.) Morong, *S. emersum* Rehmann, and *S. eurycarpum* Engelm.) occur in the region that occupy wet shores or shallow waters. These emergent wetland species (excluded here), however, are generally stout, with mostly stiff, erect, and emersed stems, and emersed, distinctly keeled leaves.

KEY TO SPECIES

1. Flowering stems with 1 cluster of male flowers; fruit clusters 5–12 mm in diam.; beak of fruit 0.5–1.5 mm long... *S. natans*
1. Flowering stems with 3+ clusters of male flowers; fruit clusters 15–25 mm in diam.; beak of fruit 1.5–3.5 mm long..2
   2. Flowering stems branched; beak of mature fruit flattened and strongly curved; leaves 4–10 mm wide and flat..*S. fluctuans*
   2. Flowering stems unbranched; beak of mature fruit generally rounded and straight; leaves 2–5 mm wide with surface slightly convex... *S. angustifolium*

S. angustifolium

NARROW-LEAVED BUR-REED — *Sparganium angustifolium* Michx.

Narrow-leaved bur-reed is a submersed, rooted perennial. Leaves are submersed, with floating ends, long (to 1 m), ribbon-like (2–5 mm wide), with a slightly rounded surface. Flowering stems are unbranched with 2–5 female flower clusters, the lowest one on a stalk arising from the axil of a leafy bract. Male flower clusters number 4–10+. Fruit clusters 15–25 mm in diam. Fruits dull, reddish to brown (3–7 mm long) with a straight or slightly curved beak 1.5–2 mm long.

HABITAT & DISTRIBUTION

Sparganium angustifolium is a native species of shallow lakes and slow streams. It is widespread in northern New England.

Note: Similar to the widespread Simple-stemmed bur-reed (*S. emersum*) which can sometimes have flaccid and floating leaves. Usually *S. emersum* has broader (6–12 mm wide), erect and emergent leaves which have a prominent keel along the lower surface of distal portions. Its fruit beak is also longer (2–6 mm). It is known to hybridize with *S. emersum* in NH and VT.

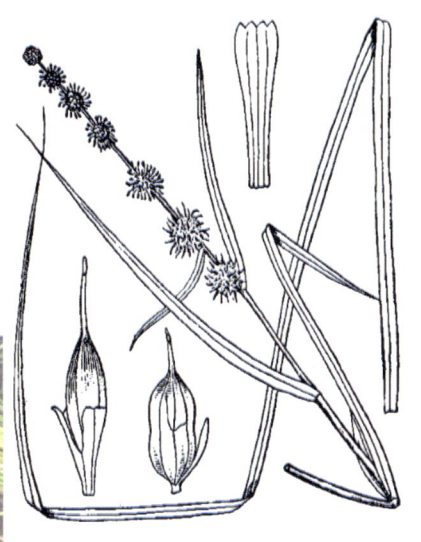

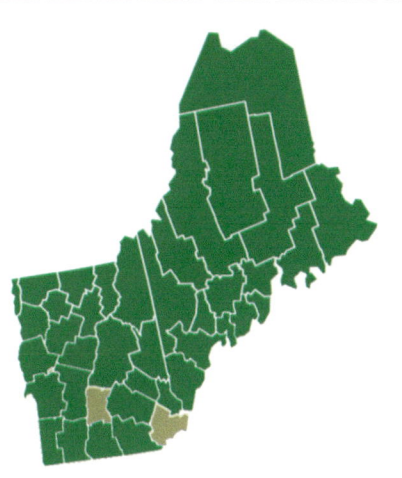

FLOATING BUR-REED — *Sparganium fluctuans* (Morong) Robinson

Floating bur-reed is a submersed, rooted perennial. Leaves are submersed, with floating ends, strap-shaped (to 1.5 m long, 3–10 mm wide), and flat (but slightly keeled at base). Flowering stems are branching with 2–4 female flower clusters and 3–6 male flower clusters. Fruit clusters 15–23 mm in diam. Fruits dull, dark reddish-brown (3–5 mm long) with a flattened, strongly curved beak 2–3.5 mm long.

HABITAT & DISTRIBUTION

Sparganium fluctuans is a native species of shallow, still lakes or slow rivers. It is uncommon in V⊤.

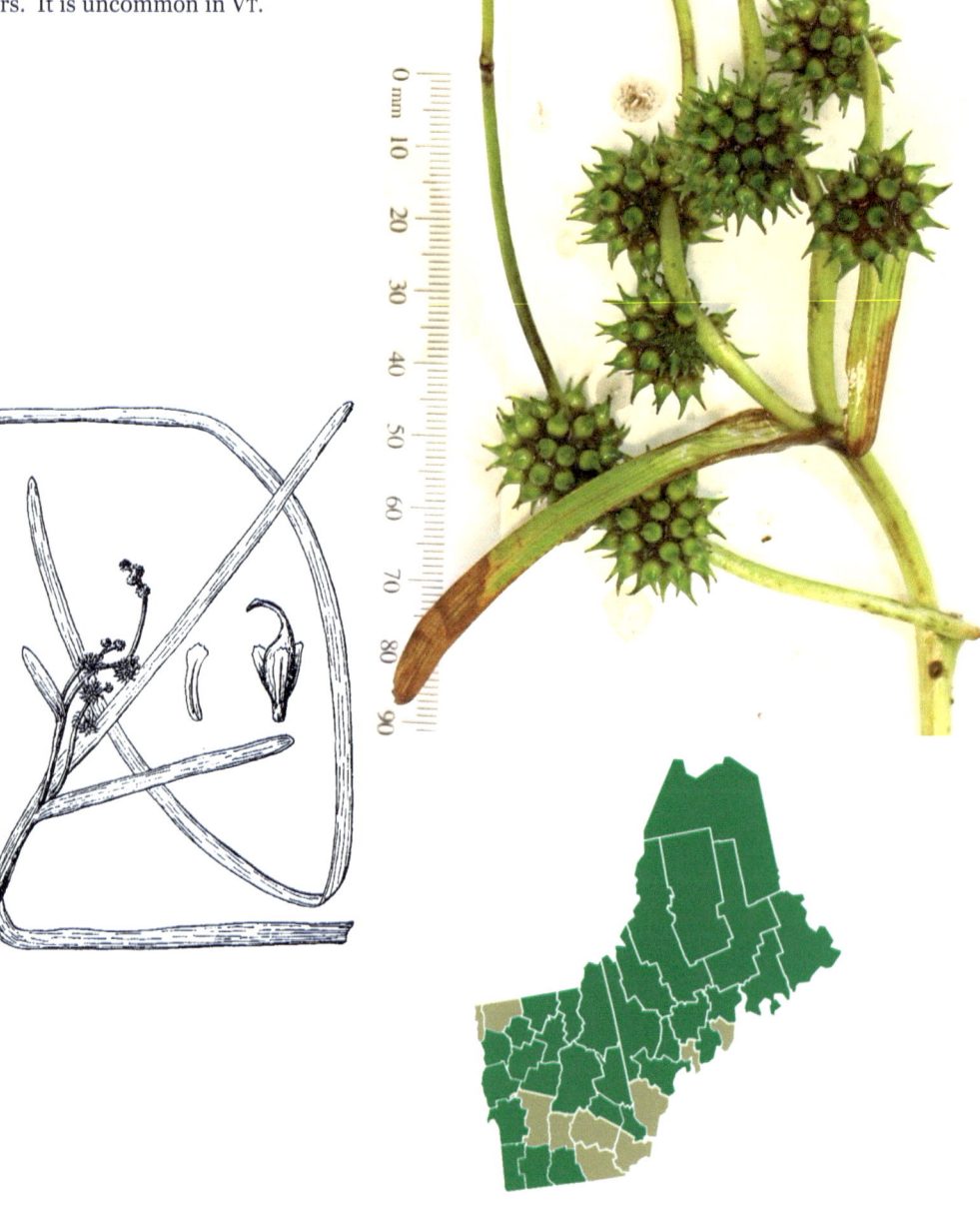

ARCTIC BUR-REED *Sparganium natans* L.

Arctic bur-reed is a submersed, rooted perennial. Leaves are submersed, with floating ends, strap-shaped (to 1.5 m long, 2–7 mm wide), and flat. Flowering stems are unbranched with 1–3 female flower clusters and only 1 male flower cluster. Fruit clusters 5–12 mm in diam. Fruits dull, greenish or brown (3–4 mm long) with a slender beak 0.5–1.5 mm long.

HABITAT & DISTRIBUTION

Sparganium natans is a native species of shallow lakes, peatlands, and slow streams. It is very rare and state-listed as *threatened* in NH and VT.

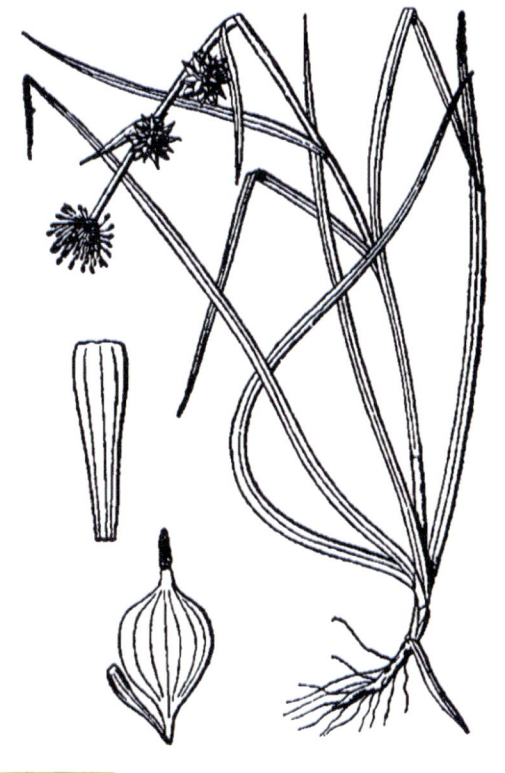

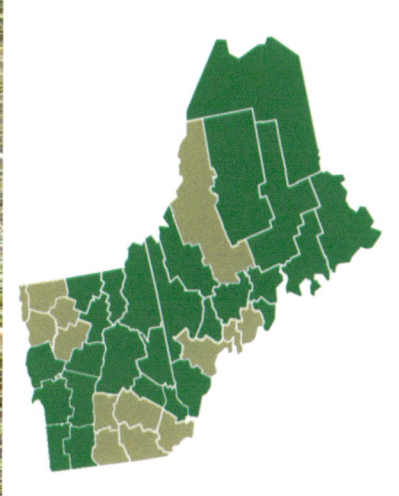

Spirodela — Araceae
Duck-meal

COMMON DUCK-MEAL — *Spirodela polyrhiza* (L.) Schleid.

Common duck-meal is a diminutive, perennial, floating on water surface. Its body, or thallus (not differentiated into stem and leaf), is small (2–10 mm long, 1.3–10 mm wide), flattened, rounded to egg-shaped, with 5–21 obscure veins and often a purple dot on top surface, and solid red-purple beneath. Slender, 7–21, unbranched roots (to 30 mm long) are clustered on lower surface extending into water. Flowers and fruits are minute, rarely seen. Turions formed late in season.

HABITAT & DISTRIBUTION
Spirodela polyrhiza is a native species of ditches, ponds, lakes, and slow rivers. It is widespread in northern New England.
Notes: Like most duckweeds, vegetative reproduction is extensive. Usually, multiple plants are connected and intermixed with other duckweeds (*Lemna* and *Wolffia*).

Stuckenia — Potamogetonaceae
False pondweed

The False pondweeds are submersed, rooted, perennials with slender, flexuous, much-branched, stems, and slender, creeping rhizomes bearing white tubers (1–2 cm long) at ends. Leaves are submersed, alternate, very narrow (≤1 mm wide), mostly 1-veined, channeled, opaque, and rigid. At the base of each leaf, there is a stipule that forms a cylindrical sheath around the stem. A portion of the stipule (greater than two-thirds of its length) is fused to base of leaf. Flowers are small, individually inconspicuous, and clustered in cylindrical spikes at ends of long (to 15 cm), flexible, stems floating on water surface. Fruits are small, hard, seed-like, achenes.

Two species occur in northern New England, both typical of alkaline, calcareous waters. The abundant underground tubers and fruits are an important food for waterfowl.

KEY TO SPECIES

1. Fruits mostly 3–4.5 mm, with a distinct (1 mm) persistent beak; stipular sheaths open with overlapping edges, leaves sharp pointed.. S. pectinatus
1. Fruits mostly 2–3 mm, beakless; stipular sheaths tightly closed, leaves minutely blunt or rounded... S. filiformis

THREAD-LEAF FALSE PONDWEED *Stuckenia filiformis* Pers.

Thread-leaf false pondweed has leaves narrowly linear (1–15 cm long, 0.2–0.5 mm wide), channeled, mid-veined, stiff, and tapered to a more or less blunt tip. Stipules (0.2–0.5 mm wide) are attached to base of leaf for most of their length, forming a tightly closed sheath (10–40 mm long) around the stem, the free tip extending as a ligule 2–20 mm long. Flowers are clustered in a cylindrical spike (1–5 cm long) of 2–6 interrupted whorls of flowers, at end of a slender, flexuous, submersed stalk (2–15 cm). Fruits are mostly 2–3 mm long, dark brown, beakless, with a low round keel.

HABITAT & DISTRIBUTION

Stuckenia filiformis is a native, northern plant of cold, shallow (<1 m), calcareous lakes, ponds, or slow streams. Restricted to northern regions, where it is rare throughout; state-listed as ***special concern*** (ME) and ***endangered*** (NH).

Note: A rare sterile hybrid involving this species [and *S. vaginata* (Turcz.) Holub.] is known from rivers in ME and VT. This hybrid (named *S.* ×*fennica* (Hagstr.) Holub) has wider leaves (to 2 mm) and loose, inflated, stipules.

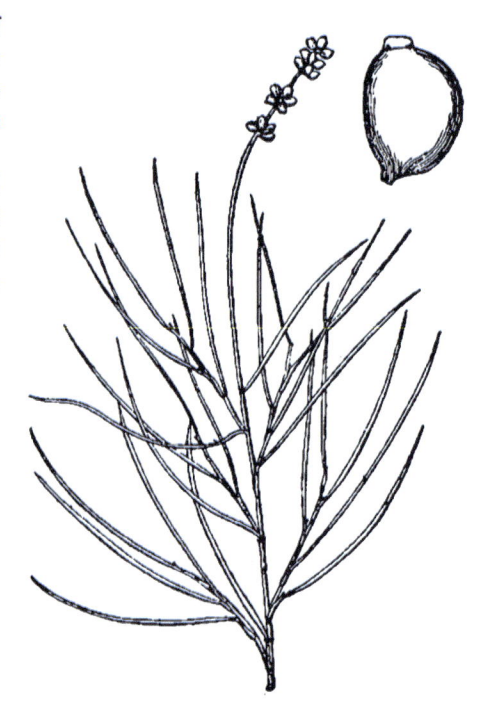

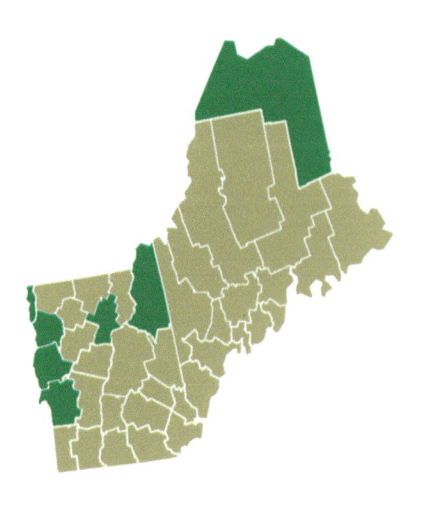

SAGO FALSE PONDWEED — *Stuckenia pectinata* (L.) Börner

Sago false pondweed has much-branched stems with numerous fan-like sets of leaves. Leaves linear (3–15 cm long, 0.2–1 mm wide), channeled, mid-veined, stiff, and tapered to a sharp-pointed tip. Stipules (1–2 cm long) are attached to base of leaf for most of their length, forming an open sheath (0.8-1.1 cm) around the stem, the free tip extending as a small ligule (0.8 mm long). Flowers are clustered in a floating, cylindrical spike (1.5–3 cm long) of 3–5 interrupted whorls of flowers, at end of a slender, flexuous, submersed stalk (3–12 cm). Fruits are mostly 3–4.5 mm long, distinctly beaked (0.5–1 mm long), brown to yellow-brown, with a round keel.

HABITAT & DISTRIBUTION

Stuckenia pectinata is a native plant of calcareous lakes, ponds, or streams, including coastal brackish waters. Widespread in northern New England, but rare and state-listed as *endangered* in NH.

Note: Colonies can form dense, bushy masses.

Subularia — Brassicaceae
Awlwort

AMERICAN AWLWORT — *Subularia aquatica* L.

American awlwort, or Water awlwort, is a small, submerged (sometimes emergent) annual with a rosette of 10–20 green leaves and bright white roots. Leaves are basal, linear (1–5 cm long), mostly cylindrical, widest at base, somewhat flattened toward base, and tapering to a sharp tip (i.e., awl-like). Flowers minute (ca. 4 mm wide) with four white petals borne on a short (2–10 cm long), branched, erect cluster of 2–12 flowers. Fruits small (2–3.5 mm long mm), oblong, inflated, and splitting in half when ripe.

HABITAT & DISTRIBUTION
Subularia aquatica is a native, diminutive, northern plant of shallow (usually <45 cm deep) lakes, pools, or slow streams. It prefers clear, acidic, low-nutrients waters with sandy-gravelly (or less often muddy) bottoms. Uncommon and scattered in ME, rare and state-listed as *endangered* in NH and *historic* in VT.

Notes: These small aquatics are usually found flowering and fruiting under water. It can grow as an emergent if stranded on wet shores. Our plants are ssp. *americana* Mulligan & Calder. Similar to American shoreweed (*Littorella americana*).

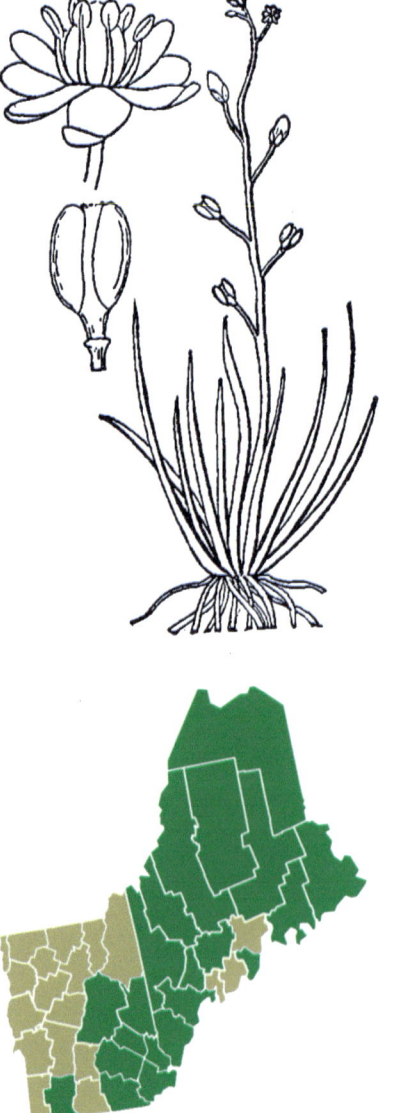

Trapa — Lythraceae
Water-chestnut

WATER-CHESTNUT — *Trapa natans* L.

Water-chestnut, or Caltrop, is a rooted, annual plant with elongated (to 1 m), flexuous stems terminating with a floating rosette (to 50 cm diam.) of triangular leaves. Leaves are of two types: floating leaves are alternate, green (4–6 cm long, 4–8 cm wide), with triangular to diamond-shaped, broadly toothed blades, and swollen and spongy stalks. Submersed leaves are alternate or opposite, pinnately divided into linear (2–18 cm long), root-like segments. Flowers are bisexual, borne singly, above water, from a floating leaf axil, with 4 white petals. Fruit is dark, dry, nut-like (2–3 cm long, 2–4.5 cm wide) with 4 stout, spreading spines.

HABITAT & DISTRIBUTION

Trapa natans is a non-native species of lakes and margins of slow rivers. It is infrequent in southern areas of northern New England where it is **invasive** and banned. Native to Eurasia, it was first introduced into North America as an ornamental plant through MA in the late 1800s. It prefers nutrient-rich, shallow (<2 m) waters and can reach high densities. This species is highly fecund, grows rapidly, and exhibits long seed dormancy. Floating rosettes can detach and float freely.

Notes: This is a highly distinctive plant with its inflated, buoyant, rosette of leaves and hard, black, spiny fruits.

Utricularia — Lentibulariaceae
Bladderwort

The Bladderworts are delicate, mostly submersed, carnivorous perennials or annuals with narrow, rootless, green stems. Plants are mostly free-floating in water or in some cases anchored to a saturated substrate by buried branches. Leaves (actually leaf-like branches) are green, mostly divided into narrowly linear segments, and bearing tiny, sac-like, animal traps (i.e., bladders). Flowers are showy, yellow, or pink-purple in a few species, and borne singly or clustered at end of an elongated, unbranched stalk (peduncle) elevated above water surface. Petals are asymmetrical, united, parted to form an upper and lower (larger) lip, and a spur (i.e., elongated hollow appendage). Fruits are rounded capsules. Many species form compact winter buds (turions) late in the season.

Ten species occur in northern New England, of which one is non-native. Without flowers present, proper species identification can be difficult.

Key to Species

1. Leaves few, unbranched and thread-like (or very narrowly linear), completely or mostly hidden in substrate; plants appearing terrestrial and only apparent when flowering; bladders few, 0.2–1.1 mm long……………………………………………………………………..……….2
1. Leaves numerous, finely forked 1-many times, all or mostly on stems floating under water, conspicuous; bladders well-developed, 1–5 mm long………………………………………………….. 3
    2. Petals purple; one flower per peduncle; scale-like bracts on peduncle paired and fused; leaves with ≥1 transverse partitions…………………………............................... *U. resupinata.*
    2. Petals yellow, 2–10 flowers per peduncle; scale-like bracts on peduncle alternate; leaves without transverse partitions………………………………………………………………….. *U. cornuta*
3. Peduncle with a whorl of spongy, inflated branches near water surface functioning as floats ……………………………………………………………………………………………………….4
3. Peduncle without strongly inflated branches…………………………………………………….5
    4. Peduncles 10–30 cm tall with 9–14 bright yellow flowers; floats 3–8 cm long and up to 8 mm wide, gradually tapered toward base……………………………………………. *U. inflata*
    4. Peduncles 3–5 cm tall with 1–7 dull yellow flowers; floats 1–4 cm long and 2–4 mm wide, with mostly parallel sides………………………………………………………………….. *U. radiata*
5. Leaves all whorled; bladders borne at tips of leaf segments and lacking external hairs; flowers pink-purple…..…………………………………………………………………..………. *U. purpurea*
5. Leaves mostly alternate or crowded at base; bladders scattered on sides of leaf segments and with whisker-like hairs at the mouth; flowers yellow……………………………...……….……..6
    6. Leaf segments distinctly flat, with a visible midrib; floral bracts lobed at base……………7
    6. Leaf segments fine and hair-like, without a midrib; floral bracts not lobed……………..….8
7. Green leaf-like branches (above substrate) lacking bladders and whitish branches (buried in substrate) bearing bladders; leaf segments spine-toothed; bladders 1.5–4 mm long… ……….. ………………………………………………………………………………………….. *U. intermedia*
7. Green leaf-like branches (above substrate) bearing some bladders; leaf segments not spine-toothed; bladders 1–2 mm…………………………………………………………………. *U. minor*

8. Plants small and simple, <15 cm long; principal leaves mostly forked 1–4 times, up to 1 cm long, the leaf segments without marginal spines; peduncles 1–2 flowered............ *U. gibba*

8. Plants stout and bushy, 30–100 cm long; principal leaves mostly forked 5–17 times, 1–6.5 cm long, the leaf segments with minute marginal spines; peduncles 2–14 flowered.......... 9

9. Leaves 1.5–6.5 cm long, coarse, usually forked 6–17 times; bladders up to 6 mm long; peduncle 6–14 flowered, with 1–3 scales below the lowest flowers; underwater flowers lacking; turions 5.5–13.5 mm thick... *U. vulgaris*

9. Leaves 1–2 cm long, extremely delicate and limp, forked 5–9 times; bladders up to 2 mm long; peduncle 2–8 flowered, without scales below the lowest flowers; bud-like, underwater flowers also borne singly on short, submersed stalks; turions 2–5 mm thick........ .... *U. geminiscapa*

HORNED BLADDERWORT — *Utricularia cornuta* Michx.

Horned bladderwort is an inconspicuous perennial. Leaves are alternate, few, slender and unbranched (needle-like), and mostly hidden within the surface of a wet soil substrate. Leaves bear few, tiny (0.2–1.1 mm long) bladders also buried in the substrate. Flowers are bright yellow, very fragrant, each with a short stalk (1–2 mm long), a long (7–14 mm) spur pointed downward, and a lower lip 9–16 mm long. Flowers are crowded 2–6 on an upright, stout peduncle (10–25 cm tall) with a few, widely spaced, alternate, scale-like bracts. Capsules are 4–5 mm long. No turions are produced.

HABITAT & DISTRIBUTION

Utricularia cornuta is a native species of bogs, pond shores, and river margins. It is widespread in northern New England, more commonly of acidic waters of boggy pools.

Note: One of the so-called "terrestrial" species of bladderwort it is only apparent when flowering. Plants can be abundant, and form thin, tangled mats on a substrate which can dislodge and float.

MIXED BLADDERWORT *Utricularia geminiscapa* Benj.

Mixed bladderwort is a submersed, bushy, perennial with slender, elongated stems. Leaves are alternate, delicate (1–2 cm long), usually forked 5–9 times into limp, hair-like segments, with a few minute, marginal spines. Leaves bear numerous bladders (1–2 mm long). Flowers are yellow, with lower lip 5–8 mm long, and borne 2–8 on an upright peduncle (5–15 cm tall). In late summer, bud-like, petal-less flowers are also borne singly at the tips of short (2.5–7.5 mm long), submersed stems. Turions are small (2–5 mm wide), loose, green, and lacking hairs.

HABITAT & DISTRIBUTION

Utricularia geminiscapa is a native species of, usually shallow, acidic, peatlands, pools, ponds, lakes, and streams. It is widespread but less common westward in northern New England.

Note: Similar to smaller plants of Common bladderwort (*U. vulgaris*) but overall reduced and finer, with submersed flowers, fewer emersed flowers, and smaller turions.

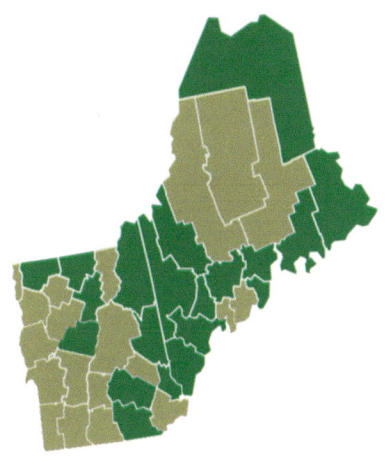

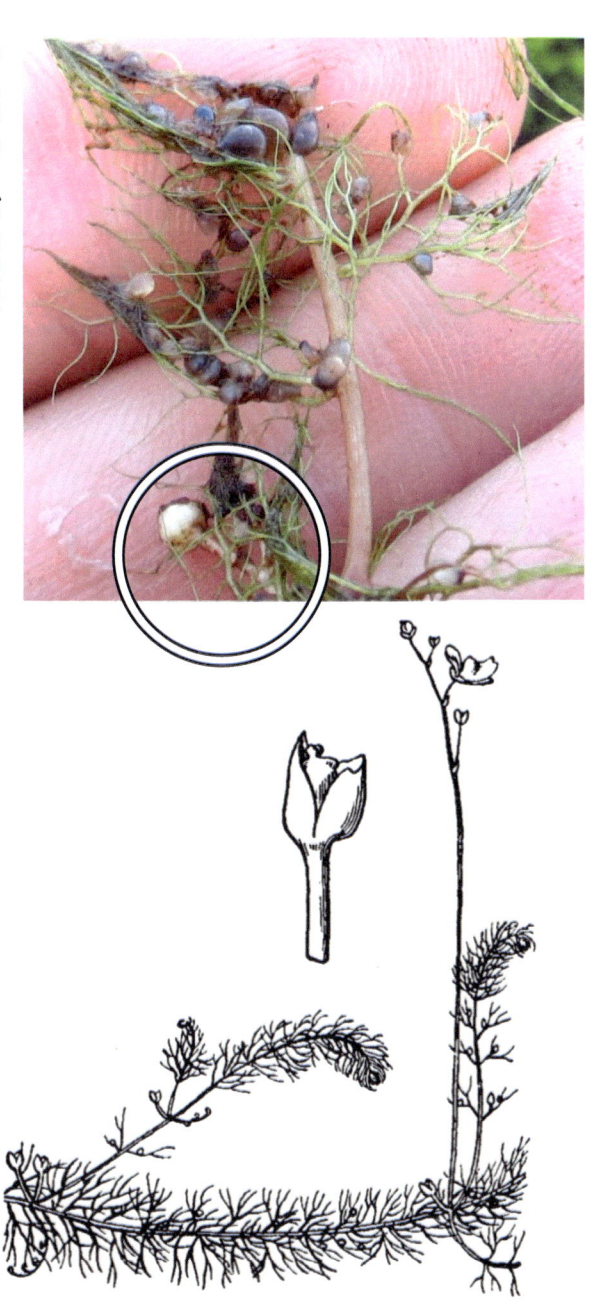

191

CREEPING BLADDERWORT *Utricularia gibba* L.

Creeping bladderwort is a delicate, submersed (or creeping), annual or perennial, with slender, short (to 10 cm long), intertwined, green stems. Leaves are alternate, short (rarely >5 mm long), mostly forked 1–2 times, and bear few, small (1 mm), bladders on sides. Flowers are yellow, with a blunt spur about half the length of the lower lip (5–6 mm long) and borne 1–2 on an upright peduncle (5–10 cm tall). Capsules are 2–3 mm long. Turions are not produced.

HABITAT & DISTRIBUTION

Utricularia gibba is a native species of lakes and ponds. It is widespread but infrequent in northern New England.

Notes: Commonly found as free-floating, intricately tangled, masses, but flowering is usually limited to plants in very shallow water (2–3 cm deep) or those stranded on wet, exposed shores.

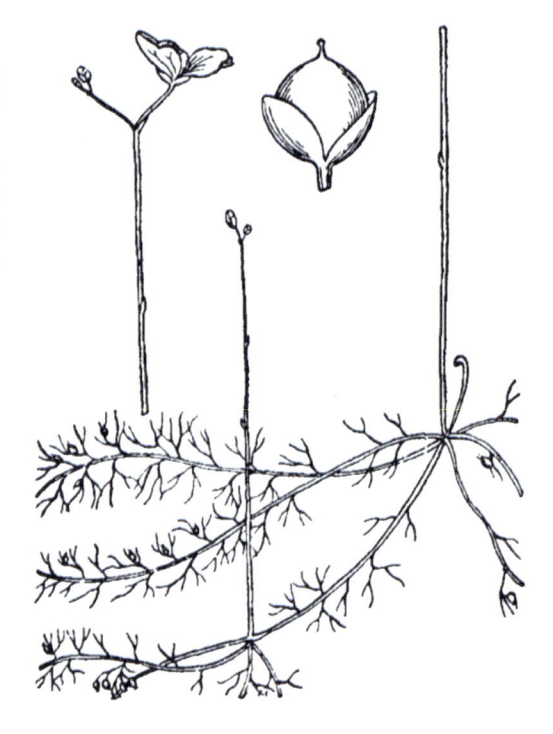

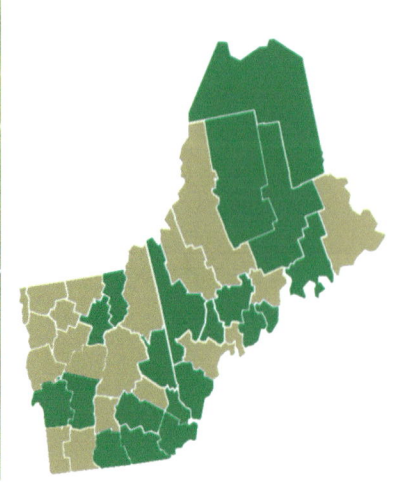

SWOLLEN BLADDERWORT *Utricularia inflata* Walt.

Swollen bladderwort is a submersed, perennial with slender, elongated stems. Submersed leaves (to 18 cm long) are alternate, bushy, repeatedly forked into thread-like segments bearing bladders on the sides. Uppermost leaves are whorled (5–10) around base of peduncle and noticeably inflated, floating at water surface. Floats (3–8 cm long, to 8 mm wide) are spongy, gradually tapered toward peduncle, with branching along half its length. Flowers are bright yellow, stalked (10–35 mm long), spur about half the length of the lower lip (20–25 mm long), and borne 9–14 (usually 10–11) on an upright peduncle (10–30 cm tall).

HABITAT & DISTRIBUTION

Utricularia inflata is a non-native species of shallow, clear, lakes and ponds. It is **introduced** and uncommon into ME. This species can reproduce vegetatively by stem fragmentation and tuber formation and has potential to become weedy.

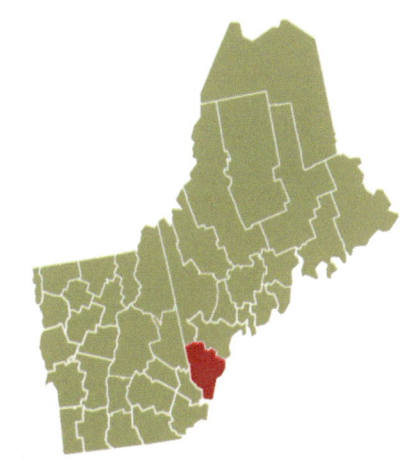

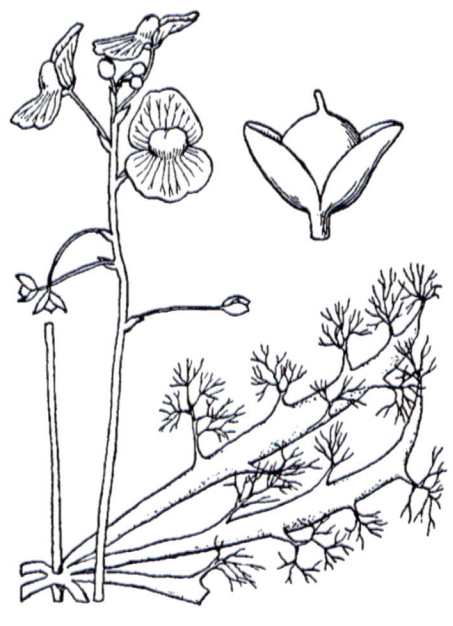

FLAT-LEAVED BLADDERWORT *Utricularia intermedia* Hayne

Flat-leaved bladderwort is a delicate, submersed (or creeping on substrate under water), perennial with slender stems. Leaves (above substrate) are alternate, green, fan-like, forked into strongly flattened, minutely spine-toothed, segments (5–20 mm long), lacking any bladders. Separate, whitish, leaf-like shoots (usually embedded in substrate) bear bladders (1.5–4 mm long). Flowers are bright yellow, with a long (8–12 mm) lower lip, well-developed spur (to 12 mm), and borne 2–5 on an upright peduncle (6–20 cm tall). Turions are oval (3–10 mm long).

HABITAT & DISTRIBUTION

Utricularia intermedia is a native species of shallow waters of ponds, lakes, bogs, fens, and marshes. It is widespread in northern New England.

Notes: The species is intolerant of eutrophication. Similar to Lesser bladderwort (*U. minor*), but with green bladder-less leaves and larger, bright yellow flowers.

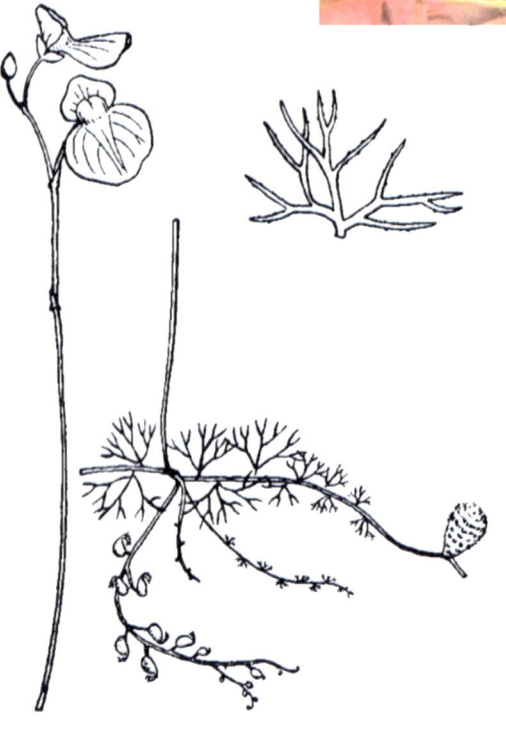

195

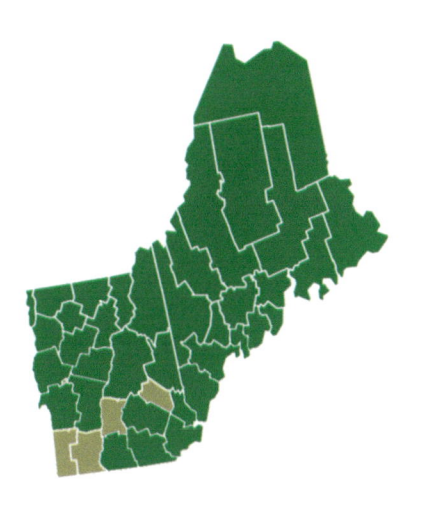

LESSER BLADDERWORT *Utricularia minor* L.

Lesser bladderwort is a delicate, submersed (or creeping on wet substrate), perennial with slender stems. Leaves are alternate, numerous, fan-like, and usually forked 3–4 times into distinctly flattened segments (3–10 mm long). Green leaves bear few, small (1–2 mm long) bladders while whitish leaves embedded in substrate bear most bladders. Flowers are pale yellow, small (petals 6–8 mm long), with an obscure (almost absent) spur, and borne 2–8 on an upright peduncle (4–15 cm tall). Turions are tiny (<3 mm), reddish-green, and rounded.

HABITAT & DISTRIBUTION

Utricularia minor is a diminutive, native species of shallow (<0.5 m deep), typically nutrient-poor, water of ponds, fens, and rivers. It is uncommon in northern New England.
Notes: Plants are rarely found in flower, so they can be easily overlooked. Flattened aspect of leaves easily seen under hand lens. Similar to Flat-leaved bladderwort (*U. intermedia*), but with green leaves bearing bladders and smaller, pale flowers.

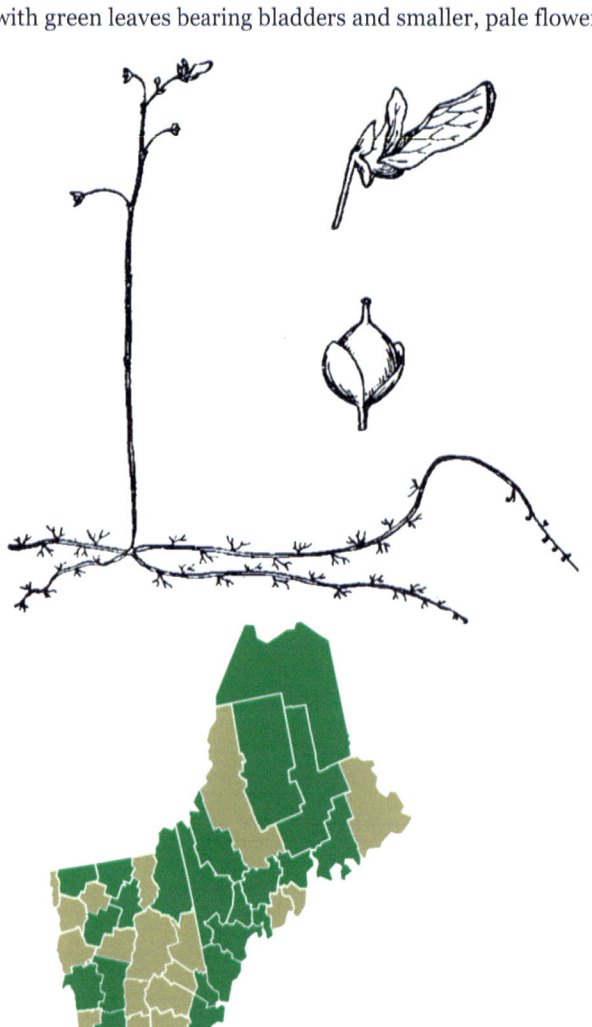

PURPLE BLADDERWORT *Utricularia purpurea* Walt.

Purple bladderwort is a submersed, perennial, with slender, elongated (to 1 m), much-branched stems. Leaves are in whorls of 5–7, conspicuous, stalked (1–1.5 cm long), forked several times into thread-like segments that bear bladders at their tips. Bladders are without external guide hairs. Flowers are pink-purple, with a blunt spur (2–5 mm long) pressed against the yellow-blotched lower lip, and borne 2–5 on an upright peduncle (5–15 cm tall). Turions are not produced.

HABITAT & DISTRIBUTION

Utricularia pupurea is a native species of lakes, ponds, and slow streams. It is widespread and common except in VT.

Notes: Plants are free-floating and when conditions are optimal often become abundant. This species is found mostly in acidic, low nutrient waters and considered a good indicator species of phosphorous-poor water bodies.

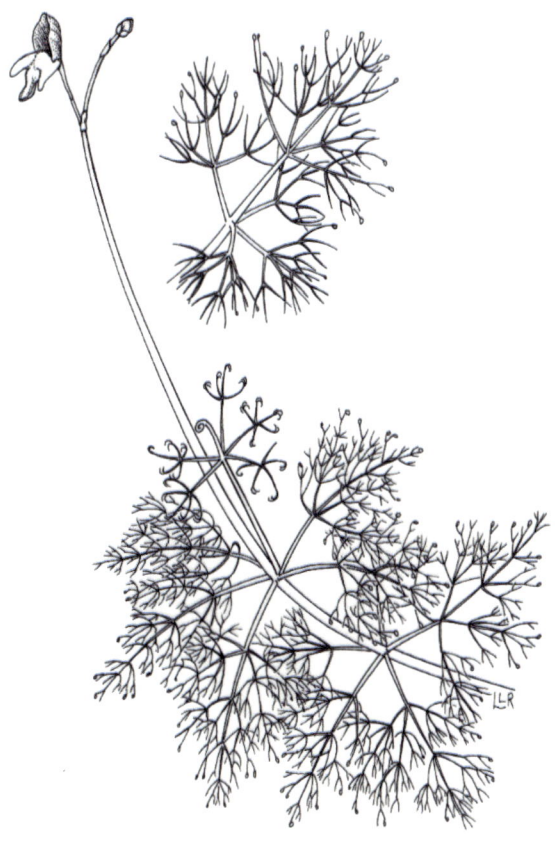

FLOATING BLADDERWORT *Utricularia radiata* Small

Floating bladderwort is a submersed, annual with slender, elongated stems. Submersed leaves (10–40 mm long) are alternate, repeatedly forked into thread-like segments, and bear many bladders on the sides. Uppermost leaves are whorled (4–7) around base of peduncle and noticeably inflated, floating at water surface. Floats (1–4 cm long, 2–4 mm wide) are spongy, mostly parallel-sided, with branching limited to tips. Flowers are dull yellow, stalked (10–20 mm long), spur pressed against the lower lip (8–10 mm long), and borne 1–7 (mostly 3–4) on an upright peduncle (3–5 cm tall). Turions are small (1 mm wide), roundish, and green.

HABITAT & DISTRIBUTION

Utricularia radiata is a native species of shallow, mostly acidic, ponds and lakes. It is uncommon in most of northern New England.

Notes: Similar to the more robust and bushier Swollen bladderwort (*U. inflata*) but with less densely branched leaves, smaller floats, shorter peduncles, and fewer flowers. Difficult to identify without floats and flowers.

RESUPINATE BLADDERWORT — *Utricularia resupinata* B.D. Greene

Resupinate bladderwort is an inconspicuous, delicate, perennial with horizontal, whitish, stems just below soil surface. Leaves are alternate, needle-like, unbranched, up to 3 cm long, with several transverse partitions. Leaves bear few, tiny (0.2–1.1 mm long) bladders often buried in the substrate. Flowers (1 cm) are purple, tipped backwards and facing upward, and borne singly at tip of an upright peduncle (2–10 cm tall) with a pair of scale-like bracts fused around peduncle. No turions are produced.

HABITAT & DISTRIBUTION

Utricularia resupinata is a native species occurring usually in sandy, very shallow (<15 cm), nutrient-poor ponds or lakes. It is uncommon; rare and state-listed as *endangered* (NH) and *threatened* (VT).

Note: One of the so-called "terrestrial" species of bladderwort, it is only apparent when flowering, which is sporadic in our region. It can form tangled mats in deep water which can dislodge and float.

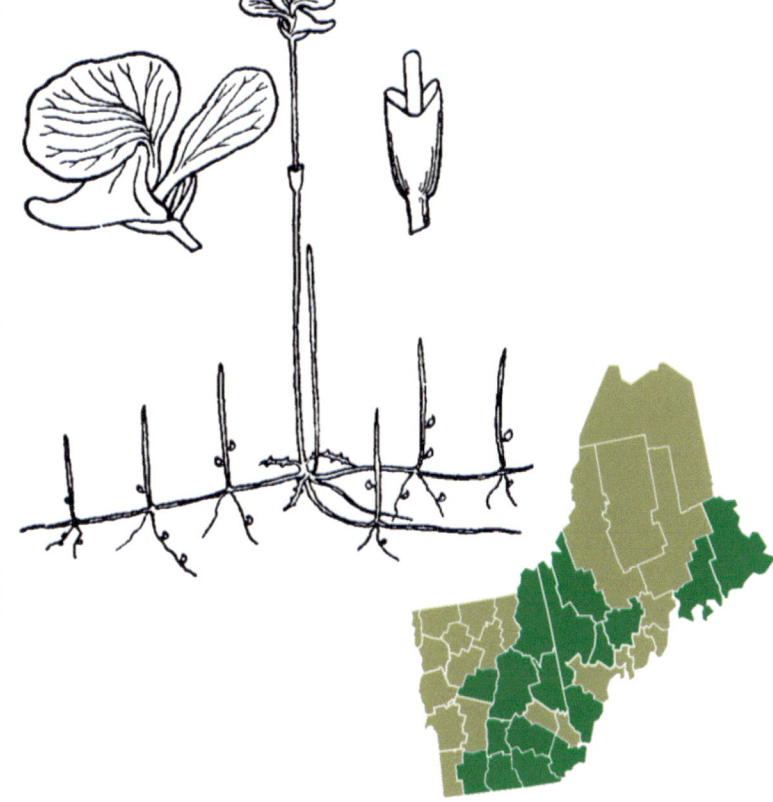

COMMON BLADDERWORT   *Utricularia vulgaris* L.

Common bladderwort, or Greater bladderwort, is a submersed, bushy, perennial with slender, elongated stems (30–100 cm long). Leaves are alternate, numerous, course (1.5–6.5 cm long), forked usually 6–17 times (and slightly zigzagged) into narrow segments, with minute marginal spines. Leaves bear numerous, conspicuous, bladders (1–6 mm long). Flowers are yellow, spur tip curved upward, lower lip 1–2 cm long, and borne 6–14 on an upright peduncle (6–20 cm tall). Turions are large (1–2 cm long, 0.5–1.5 cm wide), rounded or elongated, abundantly haired, often mucilaginous, and green.

HABITAT & DISTRIBUTION

Utricularia vulgaris (= *U. macrorhiza* LeConte) is a native species of ponds, lakes, and streams. It is common and widespread in northern New England.

Notes: This species, suspended and free-floating in water, has broad ecological tolerances (e.g., indifferent to water chemistry, flow, and depth) and is usually abundant. After pollination, flower stalks recurve and bend so fruit development takes place under water.

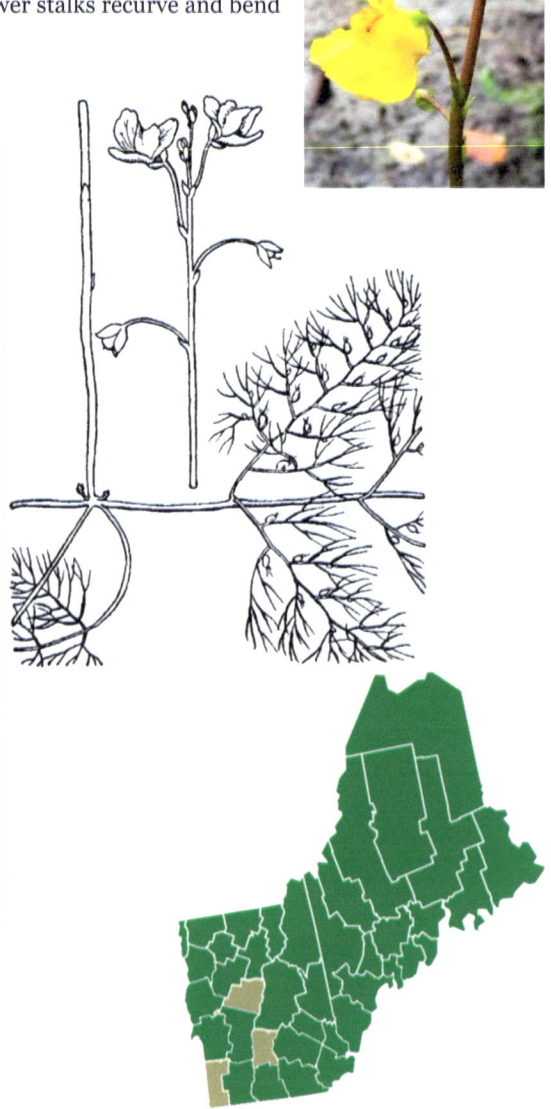

Vallisneria — Hydrocharitaceae
Tape-grass

AMERICAN EEL-GRASS — *Valisneria americana* Michx.

American eel-grass, or Water-celery, is a rooted, perennial herb with long, submersed, ribbon-like leaves clustered along stolons. Leaves are basal, submersed (upper portions may be floating), flat, long (10–110 cm long, 0.5–1.5 cm wide), bluntly pointed, bright green, and somewhat translucent. The leaf midvein has 4–5 rows of lacunae on each side, giving the blade a 3-zone appearance (light-colored middle zone bordered by darker zone on sides). Flowers are unisexual (on separate plants) and small (2 mm wide) with transparent petals. Fruit is a cylindrical (50–120 mm long) capsule borne on end of an elongated, coiled stalk.

HABITAT & DISTRIBUTION
Vallisneria americana is a native species of fresh to brackish slow rivers, streams, or lakes. It is common and widespread in northern New England.
Notes: It is a colony-forming plant and a valuable waterfowl food. Numerous male flowers detach from the plant and float to surface while single female flowers are projected to surface tethered by a long stalk. After pollination, the stalk coils and pulls the fruit underwater. It can appear very similar to non-flowering plants of *Sparganium* in deep waters.

202

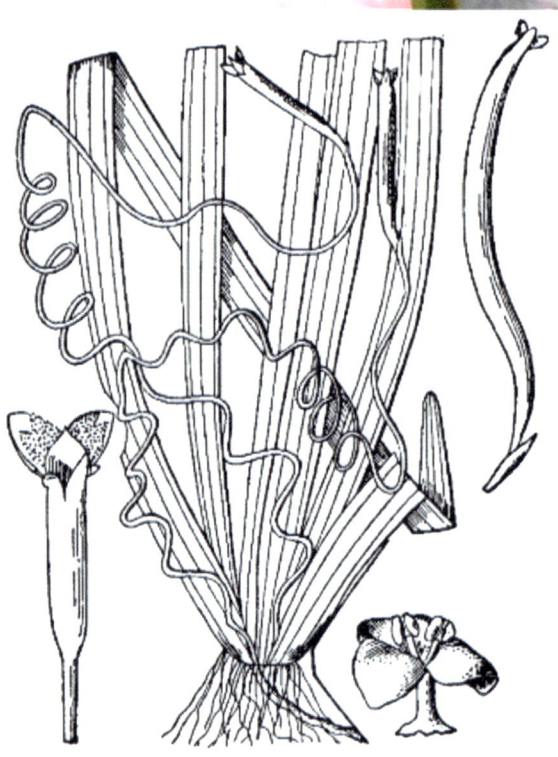

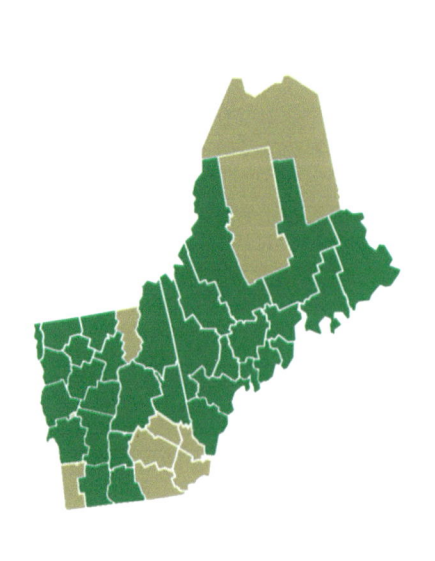

Wolffia — Araceae
Water-meal

The Water-meals are tiny (<1.5 mm long), green, rootless, perennials floating on surface of quiet waters. Plant body (called a thallus) is greatly reduced (not differentiated into stem and leaf), spherical, oval, or boat-shaped (i.e., flattened above and rounded below). Usually, two plants are connected. Flower and fruit are tiny, produced on upper surface, and rarely seen.

Three species occur in northern New England. Identification is best when plants are fresh. Individuals reproduce vegetatively quickly and are usually very numerous, often intermixed with other duckweeds (*Lemna* and *Spirodela*). This genus contains the smallest known angiosperm species in the world.

KEY TO SPECIES

1. Plants spherical, rounded in outline, largest individuals 0.5–0.8 (–1.2) mm wide .... *W. columbiana*
1. Plants oblong, longer than wide (or shaped somewhat like segments of an orange), seldom over 0.5 mm wide by 1 mm long, upper surface flattened 2
    2. Plants rounded at the ends (viewing from above); upper surface with a prominent, central, raised point in mature plant body .... *W. brasiliensis*
    2. Plants pointed at the ends; upper surface lacking a raised point .... *W. borealis*

NORTHERN WATER-MEAL *Wolffia borealis* (Engelm.) Landolt

Northern water-meal is a tiny, rootless, free-floating perennial. Plants are oblong (0.7–1.5 mm long), longer than wide, and pointed at the ends with a minute upward bend. The upper surface is flattened, smooth, without a central, raised point.

HABITAT & DISTRIBUTION
Wolffia borealis is a native plant of nutrient-rich, quiet ponds and sluggish rivers. It is uncommon in northern New England.

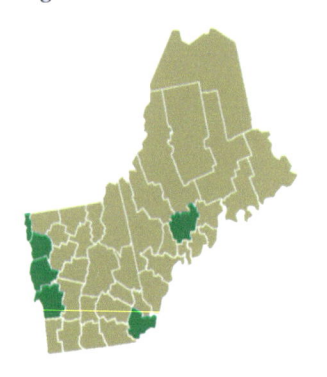

BRAZILIAN WATER-MEAL *Wolffia brasiliensis* Weddell

Brazilian water-meal is a tiny, rootless, free-floating perennial. Plants are oblong (0.5–1.6 mm long), longer than wide, and rounded at the ends. The upper surface is flattened marginally but raised in center with a prominent point.

HABITAT & DISTRIBUTION
Wolffia brasiliensis is a native plant of nutrient-rich, quiet ponds and sluggish rivers. It is rare and state-listed as *special concern* in ME.

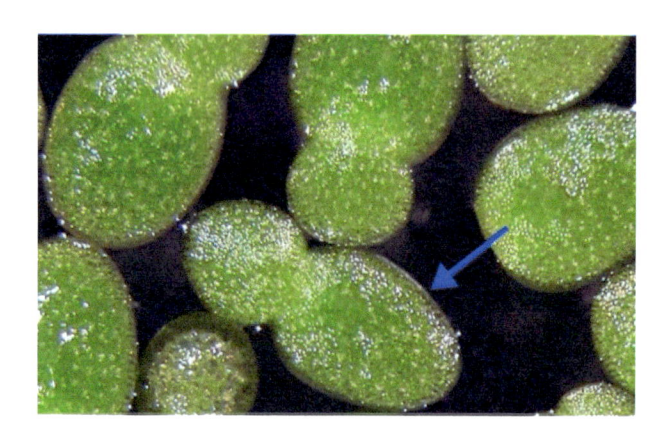

204

COLUMBIAN WATER-MEAL *Wolffia columbiana* Karst.

Columbian water-meal is a tiny, rootless, perennial, free-floating at water surface. Plants are spherical, rounded in outline. Largest individuals are 0.5–0.8 (–1.2) mm wide.

HABITAT & DISTRIBUTION

Wolffia columbiana is a native plant of nutrient-rich, quiet ponds and sluggish rivers. It is uncommon; rare and state-listed as *special concern* in ME.

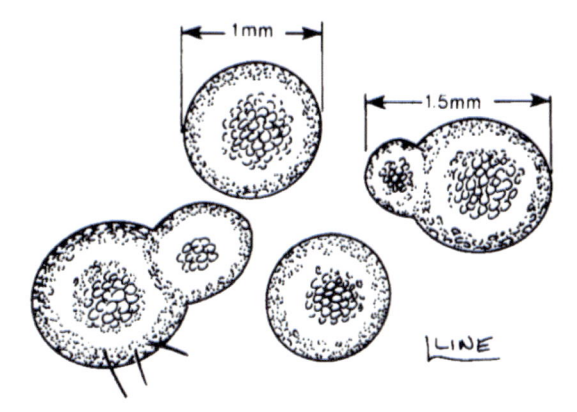

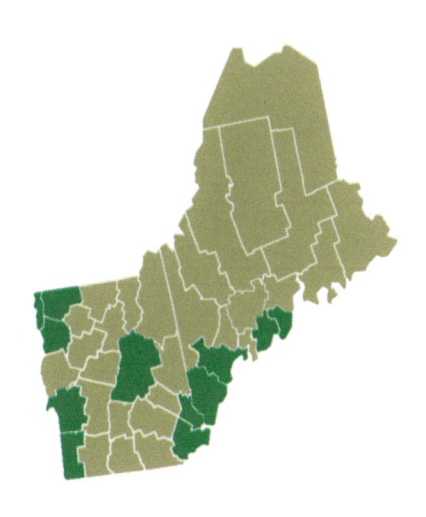

Zannichellia — Potamogetonaceae
Horned-pondweed

HORNED-PONDWEED — *Zannichellia palustris* L.

Horned-pondweed is a delicate, rooted, annual plant, entirely submersed, with many, thin (0.5 mm wide) branching stems. Leaves are opposite (possibly obscured with smaller leaves in axils), green, linear (3.5–4 cm long, 0.2–1 mm wide) with sheathing stipules at their base, and pointed at tips. Flowers are unisexual and borne underwater in leaf axils. Fruit a cluster (2–5) of short-stalked (1 mm), flattened, achenes (2–3 mm long, 1 mm wide), slightly toothed on outer edge, with narrow extension (2 mm) at tip.

HABITAT & DISTRIBUTION
Zannichellia palustris is a native plant of freshwater lakes and tidal brackish waters. It is most common in waters near the coast but can be found more inland.

Note: Sometimes confused with Wigeon-grass (*Ruppia maritima*).

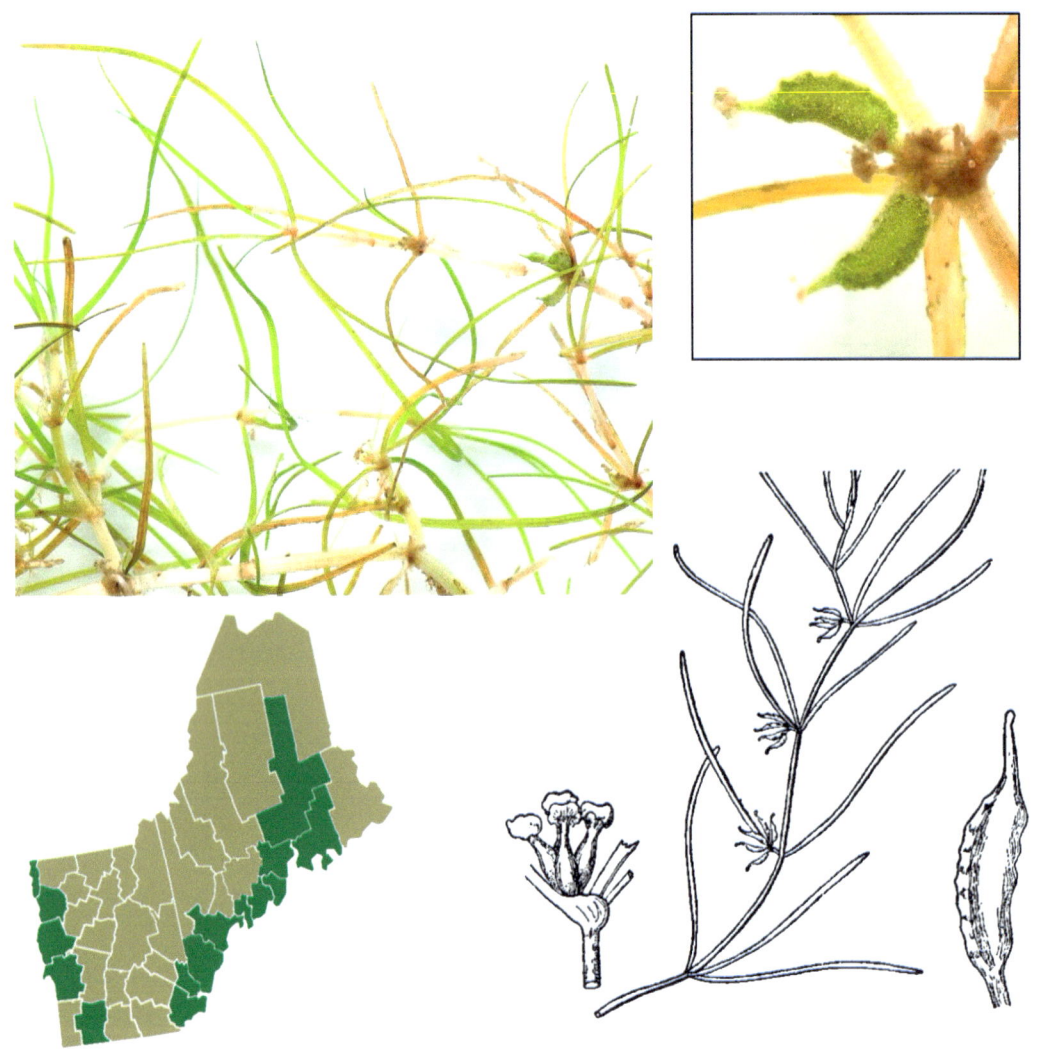

Zostera — Zosteraceae
Eel-grass

EEL-GRASS   *Zostera marina* L.

Eel-grass, or Sea-wrack, is a grass-like, completely submersed, marine perennial with rooted, creeping rhizomes (2–6 mm thick) and flattened upright branches. Leaves are alternate, dark green, and ribbon-like (30–100 cm long, 2–12 mm wide), round-tipped, and sheathing the stem at base. Flowers are minute, submersed, unisexual (on same plant), lacking petals, in rows, enclosed within a leaf-like sheath (4–8 cm long) at end of a short (2–3 cm) stalk. Fruit a fleshy, oblong (2–5 mm long), beaked drupe.

HABITAT & DISTRIBUTION
Zostera marina is a native species most common in deeper subtidal (or less frequently in exposed intertidal areas) saline waters of sheltered ocean bays and inlets. It is common along coastal areas.

Notes: This is the only submersed flowering plant of marine waters in our region. It can form vast, dense colonies or eel-grass "beds". As a sea grass, it is an extremely vital component of estuaries regarding sediment stabilization, nitrogen-fixation, animal food, shelter, and nurseries.

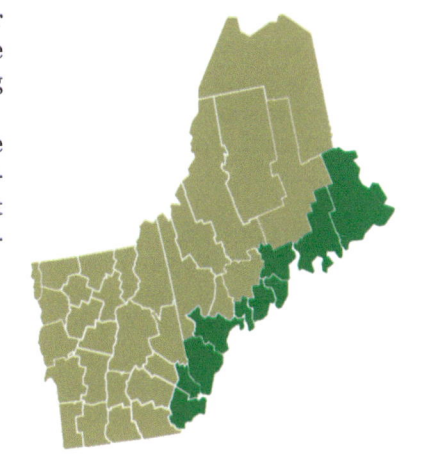

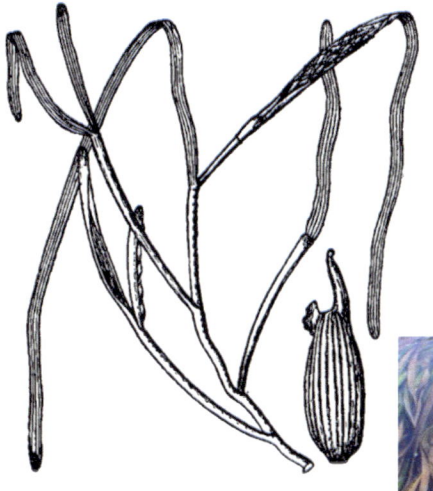

REFERENCES

Block, T.A. and A.F. Rhoads. 2011. Aquatic Plants of Pennsylvania: A Complete Reference Guide. University of Pennsylvania Press, Philadelphia.

Crow, G.E. and C.B. Hellquist. 2000. Aquatic and Wetland Plants of Northeastern North America, Vols. 1-2. University of Wisconsin Press, Madison.

Flora of North America Editorial Committee, eds. 1993+. Flora of North America North of Mexico. 21+ vols. New York and Oxford.

Gilman, A.V. 2015. New Flora of Vermont. New York Bot. Gard. Mem. 110: 1–614. New York Botanical Garden Press, Bronx.

Gleason, H.A. and A. Cronquist. 1991. Manual of Vascular Plants of Northeastern United States and Adjacent Canada. New York Botanical Garden, Bronx, New York.

Haines, A. 2011. New England Wild Flower Society's Flora Novae Angliae. Yale University Press, New Haven, Connecticut.

Hutchinson, G.E. 1975. A Treatise on Limnology: Limnological Botany, Vol. 3. John Wiley & Sons, New York.

Les, D.H. 2017. Aquatic Dicotyledons of North America: Ecology, Life History, and Systematics. CRC Press, New York.

Les, D.H. 2020. Aquatic Monocotyledons of North America: Ecology, Life History, and Systematics. CRC Press, New York.

Les, D.H., E.L. Peredo, U.M. King, L.K. Benoit, N.P. Tippery, C.J. Ball, and R.K. Shannon. 2015. Through thick and thin: Cryptic sympatric speciation in the submersed genus *Najas* (Hydrocharitaceae). Molecular Phylogenetics and Evolution 82: 15-30.

Mattson, M.D., P.J. Godfrey, M.F. Walk, P.A. Kerr, and O.T. Zajicek. 1992. Regional chemistry of lakes in Massachusetts. Water Resources Bull. 28: 1045-1056.

Norton, S.A., D.F. Brakke, J.S. Kahl, and T.A. Haines. 1989. Major influences on lake water chemistry in Maine. Maine Geol. Surv. 5: 109-124.

Thompson, E.H. and E.R Sorenson. 2000. Wetland, woodland, wildland. Vermont Department of Fish and Wildlife and The Nature Conservancy.

ILLUSTRATION & PHOTO CREDITS

Beyond my own photographs and illustrations, illustrations and photographs were utilized either from the public domain, with permission from authors, or according to various Creative Commons licenses. The vast majority of line drawings originated from Britton, N.L., and A. Brown. 1913. *An illustrated flora of the northern United States, Canada and the British Possessions. 3 vols.* Charles Scribner's Sons, New York (via USDA-NRCS PLANTS Database).

Photographs or drawings not listed with a credit below are either in the public domain or represent works of my own. For frequently cited sources, the following acronyms are used for brevity:

APP   S.L. Winterton, *Aquarium and Pond Plants of the World*, Edition 3 (https://idtools.org/id/appw/)
CBH   © C. Barre Hellquist; Used with permission.
MIN   © Katy Chayka/Peter M. Dziuk, www.minnesotawildflowers.info. Used with permission.
UFL   © Univ. Florida, Center for Aquatic and Invasive Plants. Used with permission.
WF   USDA-NRCS PLANTS Database / USDA NRCS. *Wetland flora: Field office illustrated guide to plant species.* USDA Natural Resources Conservation Service.

Cover: E. McCarty, File:Green lily pads on water (Unsplash).jpg - Wikimedia Commons (CCO-1.0)
Page 8, Figure 1, a–e: **UFL**
Page 9, Figure 2, a.: **UFL**; b.: **CBH**; d.: **UFL**; g.: **UFL**
Page 20, Photo: (whorl of leaves) S. Lefnaer, https://commons.wikimedia.org/wiki/File:Aldrovanda_vesiculosa_sl23.jpg (CC-BY-SA-4.0).
Page 21, Drawing: **UFL**; Photos: K. Ozment, https://www.inaturalist.org/observations/18738845 (CCO-1.0); I. Taylar, https://commons.wikimedia.org/wiki/File:Azolla_caroliniana.jpg (CC-BY-2.0).
Page 23, Photos: R. Foster, https://www.inaturalist.org/observations/92470784 (CC-BY-4.0); https://www.inaturalist.org/observations/31649871 (CC-BY-4.0).
Page 24, Photos: (hand) B. Michalek, https://www.inaturalist.org/observations/30476343 (CCO-1.0); (leaf): **APP**
Page 25, Photo: (bottom) S. Hollerich Giles, https://www.inaturalist.org/photos/8876308 (CCO-1.0)
Page 26, Photos: L.J. Mehrhoff, https://commons.wikimedia.org/wiki/File:Cabomba_caroliniana_5447098.jpg (CC-BY-3.0); https://www.invasive.org/browse/detail.cfm?imgnum=5447111 (CC-BY-3.0).
Page 27, Drawing: **WF**
Page 28, Photos: (*C. palustris*) R. Durand, https://www.inaturalist.org/observations/68629881 (CC-BY-4.0); (*C. stagnalis*) dannali, https://www.inaturalist.org/observations/64844397 (CC0-1.0).
Page 29, Photo: M. Bowser, https://www.inaturalist.org/observations/15232773 (CC-BY-4.0).
Page 30, Photos: (habit) L. Allain, U.S. Geological Survey; (rosette) S. Kieschnick, https://www.inaturalist.org/observations/71669771 (CC-BY-4.0).
Page 31, Drawing: Coste and Flahault, *Flore descriptive et illustrée de la France, de la Corse et des contrées limitrophes*, 1901-1906 (Public Domain); Photo: R. Routledge, https://www.invasive.org/browse/detail.cfm?imgnum=5541152 (CC-BY-3.0)
Page 32, Drawing: Coste and Flahault, *Flore descriptive et illustrée de la France, de la Corse et des contrées limitrophes*, 1901-1906; Photo: L.J. Mehrhoff, https://www.invasive.org/browse/detail.cfm?imgnum=5447134 (CC-BY-3.0).
Page 33, Drawing: **UFL**
Page 34, Drawing: C.F. Reed, 1970. Selected weeds of the United States. USDA Agric. Res. Ser. Agric. Handbook 336.
Page 35, Photo: (leaf) D. Suitor, https://www.inaturalist.org/observations/84200597 (CC-BY-3.0).
Page 37, Drawing: **UFL**; Photo: Lamiot, https://commons.wikimedia.org/w/index.php?curid=42528393 (CC BY-SA 4.0)
Page 38, Drawing: **UFL**; Photos: (leaves) Amada44, https://commons.wikimedia.org/wiki/File:Eichhornia_crassipes_001.jpg (CC BY 3.0); (flowers) C. Grenier, https://www.inaturalist.org/observations/21416617 (CC0 1.0).
Page 39, Photo: R. Foster, https://www.inaturalist.org/observations/21099340 (CC BY 4.0)
Page 42, Photo: B. Knutsen, Norway
Page 43, Photo: G. Grzejszczak, https://www.inaturalist.org/observations/14359941 (CC-BY-4.0)
Page 45, Photo: N. Martineau, https://www.inaturalist.org/observations/12871902 (CC-BY-4.0).
Page 46, Photo: **MIN**
Page 47, Drawing: **UFL**; Photo: R. Martin

Page 48, Photo: C. Fischer, https://commons.wikimedia.org/wiki/File:ElodeaCanadensis.jpg (CC-BY-SA3.0)
Page 49, Photo: C. Fischer, https://commons.wikimedia.org/wiki/File:ElodeaNuttallii2.jpg (CC-BY-SA-3.0)
Page 51, Photo: Jomegat, https://commons.wikimedia.org/wiki/File:Eriocaulon_aquaticum_6884.jpg (CC-BY-SA-3.0)
Page 53, Photo: A. Harris, https://www.inaturalist.org/observations/36740921 (CC0-1.0)
Page 54, Photos: (top) MIN; (bottom) J. Grant, https://www.inaturalist.org/observations/28158863 (CC BY 4.0)
Page 56, Photo: (flower) https://www.inaturalist.org/observations/53759500 (CC0 1.0)
Page 57, Photo: https://www.inaturalist.org/observations/49741816 (CC0 1.0)
Page 58, Photo: C. Fischer, https://commons.wikimedia.org/wiki/File:HottoniaPalustrisSubmerse.jpg (CC-BY-SA-3.0)
Page 59, Photo: (Leaves) L. Allain, U.S. Geological Survey (CC0 1.0)
Page 60, Drawing: *Bilder ur Nordens Flora*, https://commons.wikimedia.org/wiki/File:139_Hottonia_palustris.jpg (CC0 1.0); Photos: (Leaves) T. Pocius, https://www.inaturalist.org/observations/74611556 (CC0 1.0); (flowers) C. Fischer, https://commons.wikimedia.org/wiki/File:HottoniaPalustrisInflorescence.jpg (CC-BY-SA-3.0).
Page 61, Drawing: **UFL**; Photos: (in hand) Darkmax, https://commons.wikimedia.org/wiki/File:Hydrilla_Verticillata_2.jpg (CC-BY-SA-3.0); (plant) https://www.inaturalist.org/observations/57792020 (CC0-1.0)
Page 63, Photos: B. Kennedy, https://www.inaturalist.org/observations/7688231 (CC-BY-SA-4.0); A. Penney, https://www.inaturalist.org/observations/1766660 (CC-BY-4.0).
Page 65, Drawing: C.A.M. Lindman, *Bilder ur Nordens Flora*; Photo: L. Gibson, https://www.inaturalist.org/observations/7595846 (CC-BY-4.0).
Page 66, Photo: J. Lee, https://upload.wikimedia.org/wikipedia/commons/b/b4/Isoetes_echinospora_%2820262801982%29.jpg (CC BY-SA 2.0)
Page 68, Photo: D. Inozemtseva, https://upload.wikimedia.org/wikipedia/commons/a/a5/Isoetes_lacustris_rivnenskyi.jpg (CC-BY-SA-4.0)
Page 70, Photos: (flowers) B. Armstrong, https://www.inaturalist.org/observations/15396926 (CC BY 4.0); (leaf) Q. Wiegersma, https://www.inaturalist.org/observations/59942968 (CC BY 4.0).
Page 71, Photo: (hand) https://www.inaturalist.org/observations/30677513 (CC BY 4.0)
Page 72, Drawing: **WF**; Photos: Tomas P., https://www.inaturalist.org/observations/23326048 (CC0-1.0); J. Brew, https://www.inaturalist.org/observations/59348611 (CC-BY-4.0).
Page 73, Drawing: **WF**; Photo: (left) C. Fischer, https://commons.wikimedia.org/wiki/Lemna_trisulca#/media/File:LemnaTrisulca.jpg (CC BY-SA 3.0)
Page 74, Drawing: **WF**
Page 76, Photo: M. Bedingfield, https://canberra.naturemapr.org/Community/Sightings/Details/3390212 (CC BY 3.0 AU)
Page 77, Photo: D. Stokholm, https://www.inaturalist.org/observations/27922820 (CC BY 4.0)
Page 80, Photo: J. Hollinger, https://commons.wikimedia.org/wiki/File:Water_Purslane_(1290675069).jpg (CC-BY-2.0)
Page 81, Photos: (top) Q. Groom, https://www.inaturalist.org/observations/8260170 (CC0 1.0); (bottom) Summit Metro Parks, https://www.inaturalist.org/observations/20945297 (CC BY 4.0)
Page 82, Photo: K. Ziarnek, https://commons.wikimedia.org/wiki/File:Marsilea_quadrifolia_kz1.jpg (CC-BY-SA 4.0).
Page 84, Photos: (*M. spicatum* plant) A. Fox, Univ. Florida, Bugwood.org (CC BY 3.0); (*M. sibiricum*) mbelitz, https://www.inaturalist.org/observations/8534085 (CC0 1.0); (*M. spicatum* flowers) Y. Danilevsky, https://www.inaturalist.org/observations/19902829 (CY BY 4.0)
Page 85, Photo: R. Foster, https://www.inaturalist.org/observations/31649880 (CC BY 4.0)
Page 87, Drawing: **UFL**
Page 89, Photo: Hyun Jung Cho, Mississippi Aquatic Plants Website, with permission.
Page 90, Photos: (left) k2018lena, https://www.inaturalist.org/observations/27162625 (CC0); (right) N. Filippova https://www.inaturalist.org/observations/19930673 (CC-BY-4.0).
Page 91, Drawing: **UFL**
Page 92, Photos: (leaf) S. Lefnaer, https://commons.wikimedia.org/wiki/File:Myriophyllum_spicatum_sl5.jpg (CC-BY-SA-4.0); (spike) S. Lefnaer, https://commons.wikimedia.org/wiki/File:Myriophyllum_spicatum_sl1.jpg (CC-BY-SA-4.0); (habit) Елена Смирнова, https://www.inaturalist.org/observations/49493657 (CC0 1.0)
Page 93, Photo: (flowers) D. McCorquodale https://www.inaturalist.org/observations/32687637 (CC-BY-4.0).
Page 94, Photos: C. Hohn, https://www.inaturalist.org/observations/12964590 (CC-BY-4.0); Y. Basov, https://www.inaturalist.org/observations/56117341 (CC-BY-4.0).
Page 95, Drawings: **UFL**
Page 96, Photo: (*N. flexilis*) B. Isaac https://www.inaturalist.org/observations/56414038 (CC0-1.0).
Page 97, Photo: (top) adapted from Les et al. (2015) Molecular Phylogenetics and Evolution 82: 15-30; (bottom) S. Johnson, https://www.inaturalist.org/observations/30429985 (CC BY 4.0)
Page 99, Photo: Show_ryu, https://commons.wikimedia.org/wiki/File:Najas_gracillima.JPG (CC-BY-SA-3.0)
Page 100, Photo: **WF**
Page 101, Drawing: (plant) Gilg and Schumann. 1900. *Das Pflanzenreich Hausschatz des Wissens*; Photo: L.J. Mehrhoff, https://www.invasive.org/browse/detail.cfm?imgnum=5446020 (CC-BY-3.0)
Page 102, Drawing: **UFL**; Photos: https://www.inaturalist.org/observations/17986413 (CC0-1.0).
Page 104, Photos: (leaves) https://www.inaturalist.org/observations/83817563 (CC0-1.0); (flower) Mfeaver, https://www.inaturalist.org/observations/77833102 (CC BY 4.0)
Page 108, Photos: (flower) https://commons.wikimedia.org/wiki/File:Nymphaea_leibergii.jpg (CC-BY-2.0); (leaves) https://commons.wikimedia.org/wiki/File:Nymphaea_leibergii5_(5097938122).jpg (CC-BY-2.0)
Page 111, Photo: (top) Gouwenaar, https://commons.wikimedia.org/wiki/File:18820180704_Dotterbloemen_Veersloot_Molensloot_langs_Pannepad.jpg

(CC-BY-SA-4.0)

Page 113, Photo: **APP**

Page 114, Photo: (right) C. Hohn, https://www.inaturalist.org/observations/13809078 (CC-BY-4.0).

Page 115, Photo: (top) D. Suitor, https://www.inaturalist.org/observations/96405335 (CC-BY-4.0); (bottom) D. Goldman, hosted by the USDA-NRCS PLANTS Database (Public Domain)

Page 117, Photos: (*P. gramineus*) J. Hollinger, https://commons.wikimedia.org/wiki/File:Potamogeton_gramineus_(3817604665).jpg (CC-BY-2.0); (*P. natans*) C. Fischer, https://commons.wikimedia.org/wiki/File:PotamogetonNatans2.jpg (CC-BY-SA-3.0); (*P. obtusifolius*) K. Ziarnek, Kenraiz, https://commons.wikimedia.org/wiki/File:Potamogeton_obtusifolius_kz01.jpg (CC-BY-SA 4.0); (*P. zosteriformis*) A. Gunnar, https://www.inaturalist.org/observations/15942868 (CC-BY-4.0).

Page 118, Drawings: (plant) **UFL**; (node & fruit) **CBH**

Page 121, Photos: (*P. perfoliatus*) C. Fischer, https://commons.wikimedia.org/wiki/Potamogeton_perfoliatus#/media/File:PotamogetonPerfoliatus3.jpg (CC-BY-SA 3.0); (*P. gemmiparus*) **CBH**.

Page 122, Photos: https://www.inaturalist.org/photos/153685177 (CC BY 4.0); É. Lacroix-Carignan, https://www.inaturalist.org/observations/68713866 (CC0 1.0)

Page 123, Photo: C. Hohn, https://www.inaturalist.org/observations/7825307 (CC0 1.0)

Page 124, Photos: (top) **MIN**; (bottom) Q. Wiegersma, https://www.inaturalist.org/observations/34427597 (CC BY 4.0)

Page 125, Drawing: van de Weyer & Schmidt, https://commons.wikimedia.org/wiki/File:Potamogeton_berchtoldii_illustration_(01).png (CC-BY-SA-3.0); photo: H. Tinguy, https://commons.wikimedia.org/wiki/File:Potamogeton_berchtoldii_plant_(01).jpg (CC-BY-SA-2.0-FR)

Page 127, Drawing: **UFL**; photos: N. Martineau, https://www.inaturalist.org/observations/15712219 (CC-BY-4.0); Q. Wiegersma, https://www.inaturalist.org/observations/94890555 (CC BY 4.0)

Page 129, Drawing: **UFL**

Page 131, Photos: (top) R. Foster, https://www.inaturalist.org/observations/20819808 (CC-BY-4.0); (middle) Q. Wiegersma, https://www.inaturalist.org/observations/33730316 (CC-BY-4.0); (bottom) C. Hohn, https://www.inaturalist.org/observations/7193394 (CC0-1.0).

Page 133, Photo: K. Peters, https://commons.wikimedia.org/wiki/File:Potamogeton_friesii.jpeg (CC BY-SA 3.0)

Page 134, Photo: **CBH**

Page 135, Photos: (left) M. Bowser, https://www.inaturalist.org/observations/3642607 (CC-BY-4.0); (right) R. Foster, https://www.inaturalist.org/observations/20170344 (CC-BY-4.0)

Page 136, Drawing: Winterringer, *Aquatic plants of Illinois*, 1966 (Public Domain); Photos: (top) **MIN**; (bottom) Lallen, https://www.inaturalist.org/observations/106403852 (CCO-1.0).

Page 137, Photo: (top) **CBH**

Page 138, Photo: (left) **MIN**

Page 139, Drawing: **UFL**

Page 140, Photos: (left)); I. Bashinskiy, https://www.inaturalist.org/observations/88367574 (CC-BY-4.0); (right) S. Lefnaer, https://commons.wikimedia.org/wiki/File:Potamogeton_natans_sl22.jpg (CC-BY-SA-4.0)

Page 141, Photo: S. lefnaer, https://commons.wikimedia.org/wiki/File:Potamogeton_nodosus_sl3.jpg (CC-BY-SA-4.0)

Page 142, Photos: (top) C. Grenier, https://www.inaturalist.org/observations/36405373 (CCO-1.0); (bottom) D. Frade, https://www.inaturalist.org/observations/27900118 (CC BY 4.0)

Page 143, Photo: Specimen v0242029WIS, Univ. of Wisconsin – Madison herbarium (CC0 1.0)

Page 144, Photos: A. Hosper NL https://www.inaturalist.org/observations/15521384 (CC-BY-4.0)

Page 145, Photos: K. Ziarnek, https://commons.wikimedia.org/wiki/File:Potamogeton_perfoliatus_kz02.jpg (CC BY-SA 4.0); S. Lefnaer https://commons.wikimedia.org/wiki/File:Potamogeton_perfoliatus_sl9.jpg (CC-BY-SA-4.0); https://commons.wikimedia.org/wiki/File:Potamogeton_perfoliatus_sl10.jpg (CC-BY-SA-4.0)

Page 146, Photos: K. Peter, https://commons.wikimedia.org/wiki/File:Potamogeton_praelongus_bluete.jpeg (CC-BY-SA-3.0-migrated); https://commons.wikimedia.org/wiki/File:Potamogeton_praelongus.jpeg (CC-BY-SA-3.0-migrated)

Page 147, Photos: (top) https://commons.wikimedia.org/wiki/File:Potamogeton_praelongus,_Llyn_Cregennen_Isaf,_Wales.jpg (CC-BY-SA 4.0); (bottom) barbarab, https://www.inaturalist.org/observations/116903 (CC-BY-4.0)

Page 149, Photo: S. Lefnaer, https://upload.wikimedia.org/wikipedia/commons/d/da/Potamogeton_pusillus_s._str._sl9.jpg (CC BY-SA 4.0)

Page 150, Drawing: **UFL**; Photo: Q. Wiegersma, https://www.inaturalist.org/observations/33154441 (CC BY 4.0)

Page 151, Photo: M. Bowser, https://www.inaturalist.org/observations/33453024 (CC BY 4.0)

Page 154, Photo: Univ. of Wisconsin-Madison Herbarium, Catalog v0244303WIS (CC0 1.0)

Page 156, Photos: M. Bowser, https://www.inaturalist.org/observations/16033431 (CC BY 4.0)

Page 157, Drawings: **UFL**

Page 159, Photos: (top) L. Clark, https://www.inaturalist.org/observations/12889835 (CC BY 4.0); (bottom) R. Watson, https://www.inaturalist.org/observations/16804436 (CC BY 4.0)

Page 160, Photos: (top) C. Fischer, https://commons.wikimedia.org/wiki/File:RanunculusTrichophyllus.jpg (CC BY SA 3.0); (bottom) R. Martin, https://www.inaturalist.org/observations/46597959 (CC0 1.0)

Page 161, Photos: (top) R. Martin, https://www.inaturalist.org/observations/7937190 (CC0-1.0); (bottom) S.A. Schmid, https://www.inaturalist.org/observations/83165894 (CC BY 4.0)

Page 162, Photos: C. Catto, https://www.inaturalist.org/observations/86176938 (CC BY 4.0)

Page 163, Drawing: M. Cilenšek.1892. *Naše škodljive rastline* (Public Domain); Photos: (left) bazwal, https://www.inaturalist.org/observations/18211442 (CC BY 4.0); (right) H. Rose, https://commons.wikimedia.org/wiki/File:Ranunculus_sceleratus_leaf1_(14758147920).jpg (CC-BY-2.0)

Page 164, Photos: A. Moro, https://commons.wikimedia.org/wiki/File:Ranunculus_trichophyllus_leaf_(08).jpg (CC-BY-SA-4.0); L. Perrie, https://www.inaturalist.org/observations/19877727 (CC BY 4.0)

Page 165, Photo: A. Harris, https://www.inaturalist.org/observations/44919906 (CC0 1.0)

Page 166, Photos: https://commons.wikimedia.org/wiki/File:Ruppia_maritima_South_Chungcheong,_South_Korea_27_Jun_2006.jpg (CC-BY-SA-3.0); J. Horn, https://www.inaturalist.org/observations/98553556 (CC-BY-4.0)

Page 168, Photos: (top) https://commons.wikimedia.org/wiki/File:Grassy_Arrowhead_(Sagittaria_graminea)_(38859446382).jpg (CC BY 2.0); (bottom) R. Foster https://www.inaturalist.org/observations/31359039 (CC BY 4.0)

Page 169, Photo: **MIN**

Page 170, Photo: R. Martin, https://www.inaturalist.org/observations/90462828 (CC0 1.0)

Page 171, Photo: Specimen 57361, Univ. of South Florida Herbarium (CC BY 3.0)

Page 173, Photo: (bottom) I. Louque, https://www.inaturalist.org/observations/5129954 (CC-BY-4.0)

Page 175, Photos: (left) **MIN**; (right) É. Lacroix-Carignan, https://www.inaturalist.org/observations/95723571 (CC0-1.0).

Page 176, Photo: Specimen UTC00003032, Intermountain Herbarium, Utah State Univ. (CC0 1.0)

Page 177, Photo: **MIN**

Page 178, Photo: H. Hillewaert, https://commons.wikimedia.org/wiki/File:Sparganium_angustifolium.jpg (CC BY-SA 4.0)

Page 179, Photos: (left) https://upload.wikimedia.org/wikipedia/commons/0/05/Sparganium_angustifolium.JPG (CC BY-SA 3.0), (right) M. Vainu,
 https://commons.wikimedia.org/wiki/Category:Sparganium_angustifolium#/media/File:Sparganium_angustifolium_in_Estonia.jpg

Page 180, Photo: R. Routledge, https://www.invasive.org/browse/detail.cfm?imgnum=5538795& (CC-BY-3.0)

Page 181, Photos: (top) J. Hollinger, https://commons.wikimedia.org/wiki/File:Small_Bur-Reed_(3818415894).jpg (CC-BY-2.0), (bottom) A. Zharkikh,
 https://commons.wikimedia.org/wiki/Category:Sparganium_natans#/media/File:2016.08.27_13.16.58_IMG_7891_-_Flickr_-_andrey_zharkikh.jpg (CC BY 2.0)

Page 182, Drawing: **UFL**; Photo: **APP**

Page 183, Photos: (left) C. Fischer, https://commons.wikimedia.org/wiki/File:PotamogetonPectinatus1.jpg (CC-BY-SA-3.0); (right) Ljaž, https://www.inaturalist.org/observations/89444297 (CC BY 4.0)

Page 184, Photo: M. Bowser, https://www.inaturalist.org/observations/17159795 (CC BY 4.0)

Page 185, Photo: (top) C. Fischer, https://commons.wikimedia.org/wiki/File:PotamogetonPectinatus2.jpg (CC-BY-SA-3.0); (bottom) S. Lefnaer,
 https://commons.wikimedia.org/wiki/File:Potamogeton_pectinatus_subsp._pectinatus_sl31.jpg (CC-BY-SA-4.0).

Page 186, Photo: M. Bowser, https://www.inaturalist.org/observations/15273693 (CC BY 4.0).

Page 188, Drawings: (*U. vulgaris*) https://commons.wikimedia.org/wiki/File:Utricularia_macrorhiza.jpg (Public Domain).

Page 189, Photos: (*U. geminiscapa*) C. Light,
 https://commons.wikimedia.org/wiki/File:Utricularia_geminiscapa_PinhookBog.jpg (CC-BY-SA-3.0); (*U. vulgaris*), Q. Wiegersma, https://www.inaturalist.org/observations/14405102 (CC BY 4.0); (*U. gibba*) cwarneke, https://www.inaturalist.org/observations/6744465 (CC0 1.0); (*U. purpurea*) Lauren, https://www.inaturalist.org/observations/81121943 (CC BY 4.0)

Page 190, Photos: (left) S. Johnson, https://www.inaturalist.org/observations/30102530 (CC BY 4.0); (right) M. Graziano.

Page 191, Photos: (left) M. Graziano; (right) R. Martin, https://www.inaturalist.org/observations/7969383 (CC0 1.0)

Page 192, Photo: (top) T. Eiben https://www.inaturalist.org/observations/29755043 (CC0 1.0)

Page 193, Photos: (left) B. O'Meara, https://www.inaturalist.org/observations/6850277 (CC0 1.0); (right) M. Graziano,.

Page 194, Photo: (top) P. Skawinski, https://commons.wikimedia.org/wiki/File:Utricularia_intermedia_plant_(03).jpg (CC-BY-SA-3.0)

Page 195, Photo: (bottom) M. Graziano, with permission

Page 196, Photo: (top) H. BELLAT, https://commons.wikimedia.org/wiki/File:Utricularia_minor_flower_(01).JPG (CC-BY-SA-2.0-FR); (bottom) Ken-ichi Ueda, https://www.inaturalist.org/observations/14710263 (CC0 1.0)

Page 197, Drawing: **UFL**; Photo: (top) **MIN**; (bottom) N. Martineau, https://www.inaturalist.org/observations/15158860 (CC BY 4.0)

Page 198, Drawing: **UFL**

Page 199, Photos: (top) **MIN**; (bottom) N. Martineau, https://www.inaturalist.org/observations/15712241 (CC BY 4.0)

Page 200, Photos: (top) N. Takebayashi,
 https://commons.wikimedia.org/wiki/File:Utricularia_macrorhiza_(29309016568).jpg (CC-BY-SA-2.0); (bottom) Rod, https://www.inaturalist.org/observations/29494616 (CC0 1.0)

Page 202, Photo: J. Fortnash, https://www.inaturalist.org/observations/63269150 (CC0-1.0).

Page 203, Photo: C. Fischer,
 https://commons.wikimedia.org/wiki/File:WolffiaArrhiza2.jpg?fastcci_from=6316484&c1=6316484&d1=15&s=200&a=fqv (CC-BY-SA-3.0)

Page 204, Photos: (top) A. Zharkikh, https://commons.wikimedia.org/wiki/File:2015.09.13_13.52.47_IMG_0449_-_Flickr__andrey_zharkikh.jpg (CC-BY-2.0); (bottom) **APP**

Page 205, Drawing: **UFL**; Photo: S..lefnaer, https://commons.wikimedia.org/wiki/File:Wolffia_columbiana_sl14.jpg (CC-BY-SA-4.0).

Page 206, Photos: C. Fischer, https://commons.wikimedia.org/w/index.php?curid=21435105 (CC BY-SA 3.0)

Page 207, Photo: C. Lansdale, https://www.inaturalist.org/observations/14362436 (CC0-1.0).

GLOSSARY

Achene: A small, dry, non-opening, one-seeded fruit

Alternate: Applied to arrangement of leaves, whereby they arise singly at each node

Annual: Plants completing their life cycle in one year or growing season

Anther: The terminal portion of a stamen; bears pollen

Axil: As in leaf axil, the angle between a stem and a leaf

Basal: Applied to arrangement of leaves, whereby they are confined to the bottom or base of the plant

Bladders: small, sac-like structures specialized to capture prey

Blade: The expanded flat portion of a leaf

Bract: A much reduced leaf at the base of a flower stalk or flower cluster

Capsule: A dry fruit comprised of multiple divisions, normally opening to release seeds

Dimorphic: Having two forms, e.g., divided submersed leaves and undivided emergent leaves

Drupe: A fleshy fruit containing a seed(s) enclosed in a hard pit

Emergent: Existing above the surface of water, surrounded by air

Emersed: Projected above water surface

Flowering stem: The stalk of a flower cluster (i.e., a peduncle)

Follicle: A dry fruit which splits only along one side at maturity

Fragmentation: Process by which a plant body separates into multiple pieces, often with each detached section capable of rooting

Globose: Spherical in shape

Hydrophobic: Water-repellant

Leaf stalk: The stalk by which a leaf is attached to stem

Linear: Line-shaped; very long and narrow, with essentially parallel sides

Midrib: The central, linear, vein that runs along the length of a leaf blade from base to apex

Midvein: The thick central vein that runs along the length of a leaf blade

Node: A place on a stem where a leaf or leaf scar is attached

Nutlet: A diminutive nut; dry, non-opening, usually one-seeded, fruit with a hard shell

Obovate: Shaped like a section through a hen's egg but with the larger end toward the top

Opposite: Applied to arrangement of leaves, buds, or scars whereby they arise in pairs positioned across from each other, one pair at each node

Ovary: The usually swollen base of the female organ which contains the eventual seeds

Palmately divided: Arrangement of leaf segments or lobes, or veins, with these diverging from a common origin at the top of a leaf stalk

Perennial: A plant that lasts for more than two years

Petal: One of the inner floral leaves in a flower, usually brightly colored, which are whorled

Phyllodial: The condition of a leaf whereby the leaf stalk is expanded and functions as a leaf, i.e., a bladeless leaf

Pinnately divided: Arrangement of leaf segments or lobes, or veins, with these displayed on either side of a central stalk or midrib, respectively

Receptacle: The expanded base of a flower

Recurved: Curved or bent downward

Rhizome: A horizontal underground stem with roots growing from it

Rosette: A circular cluster of tightly packed leaves, usually at ground level

Sepal: One of the outer floral leaves in a

flower, usually greenish, which are whorled

Septate: With internal transverse partitions or walls called septa

Sinus: The space between two lobes of a leaf

Spike: An elongated, unbranched stem bearing flowers along it that lack stalks; youngest flowers at top

Submersed: Existing under the surface of water

Spikelet: A very small spike consisting of several reduced flowers concealed behind scales; mostly in grasses and sedges

Stamen: The male, pollen-producing, organ of a flower comprising a stalk and the anther behind scales; mostly in grasses and sedges

Stipule: A leafy or linear structure at or near the base of the leaf stalk of with roots growing from it

Stolon: A horizontal aboveground stem

Stomate: A leaf pore used for gas exchange

Two-ranked: Where the leaves on a stem are arranged in two vertical columns on opposite sides of the stem, appearing as two rows of leaves

Tuber: A thickened portion of an underground stem serving as a storage organ, e.g., a potato

Tubercle: A low, round bump, e.g., on top of some *Eleocharis* fruits

Turion: A highly condensed vegetative organ specialized for overwintering, i.e., winterbud

Velum: A thin, membranous flap of tissue covering the inside surface of a *Isoëtes* sporangium.

Whorled: Applied to arrangement of leaves or flowers whereby they arise 3 or more at each node

Winterbud: see *turion*

INDEX

Aldrovanda vesiculosa	20		Tiny	105
Algae	1		*Crassula aquatica*	36
Arrowhead	167		Crowfoot	160
Common	167		Cursed	163
Estuary	174		White water	162
Grass-leaved	172		Yellow water	161
Narrow-leaved	171		Ditch-grass	166
Northern	169		Beaked	166
Quill-leaved	176		Duckweed	71
Sessile-fruited	175		Common	72
Spongy-leaved	174		Ivy-leaved	73
Awlwort	186		Minute	72
American	186		Pale	74
Water	186		Turion	72
Azolla caroliniana	21		Duck-meal	182
Beggar-ticks	22		Common	182
Beck's	22		Dwarf water lily	108
Bidens beckii	22		Eel-grass	207
Bladderwort	188		American	201
Common	200		*Egeria densa*	37
Creeping	192		*Eichhornia crassipes*	38
Flat-leaved	194		*Elatine*	39
Floating	198		*ambigua*	42
Horned	190		*americana*	40
Lesser	196		*minima*	41
Mixed	191		*triandra*	42
Purple	197		*Eleocharis*	43
Resupinate	199		*acicularis*	44
Swollen	193		*parvula*	45
Bottlebrush	57		*robbinsii*	46
Brasenia schreberi	24		Elodea	47
Bulrush	177		Brazilian	37
Water	177		*Elodea*	47
Bur-reed	178		*canadensis*	48
Arctic	181		*nuttallii*	49
Floating	180		*Eriocaulon*	50
Narrow-leaved	179		*aquaticum*	51
Simple-stemmed	179		*parkeri*	52
Buttercups	160		Fanwort	26
Cabomba caroliniana	26		Carolina	26
Callitriche	27		Featherfoil	58
heterophylla	30		American	59
palustris	31		European	60
stagnalis	32		Floating-heart	111
Caltrop	187		Little	112
Carex	1		Yellow	113
Cattails	1		*Fontinalis*	1
Ceratophyllum	33		Frog-bit	62
demersum	34		European	62
echinatum	35		*Glyceria*	53
Chara	1		*acutiflora*	53
Coontails	33		*borealis*	53
Cow lily	105		*septentrionalis*	53

Golden pert	55	Marsh	158
Grasswort	75	Mosquito fern	21
Eastern	75	Eastern	21
Gratiola aurea	55	Mosses	1
Guppy grass	100	Mudwort	76
Hatpins	50	Atlantic	76
Hedge-hyssop	55	Mud-plantain	56
Golden	55	Grass-leaved	56
Heteranthera dubia	56	*Myriophyllum*	83
Hornwort	33	*alterniflorum*	85
Common	34	*farwellii*	86
Spineless	35	*heterophyllum*	87
Horned-pondweed	206	*humile*	89
Hottonia	58	*sibiricum*	90
inflata	59	*spicatum*	91
palustris	60	*tenellum*	93
Hippuris vulgaris	57	*verticillatum*	94
Hydrilla	61	Naiad	95
Hydrilla verticillata	61	Brittle	101
Hydrocharis morsus-ranae	62	*Najas*	95
Hypericum boreale	63	*canadensis*	97
Isoëtes	64	*flexilis*	97
acadiensis	66	*gracillima*	99
echinospora	66	*guadalupensis*	100
engelmannii	67	*minor*	101
hierglyphica	67	*Nasturtium officinale*	165
lacustris	68	*Nelumbo lutea*	102
prototypus	68	*Nitella*	1
riparia	69	*Nuphar*	103
septentrionalis	69	*advena*	104
tuckermanii	69	*lutea*	103
viridimontana	69	*microphylla*	105
Juncus militaris	70	×*rubrodisca*	105
Lake cress	165	*variegata*	106
Lemna	71	*Nymphaea*	107
minor	72	*leibergii*	108
perpusilla	72	*odorata*	109
trisulca	73	*Nymphoides*	111
turionifera	72	*cordata*	112
valdiviana	74	*peltata*	113
Lilaeopsis chinense	75	*Persicaria*	114
Limosella australis	76	*amphibia*	114
Liverworts	1	*coccinea*	114
Lobelia	78	Pickerelweed	116
Water	78	Pipewort	50
Lobelia dortmanna	78	Estuary	52
Lotus	102	Parker's	52
American	102	Seven-angled	51
Water	102	*Podostemum ceratophyllum*	115
Ludwigia palustris	80	Pondweed	117
Manna grass	53	Alga	128
Eastern	53	Alpine	122
Northern	53	Berchtold's	125
Sharp-scaled	53	Big-leaved	123
Small Floating	53	Blunt-leaved	144
Mare's tail	57	Budding	134
Common	57	Clasping-leaved	145
Marsilea quadrifolia	82	Curly	129
Mermaid-weed	157	False	183
Comb-leaved	159	Fern	151

Flat-stalked	133	Water	36
Flat-stem	156	Quillwort	64
Floating	140	Acadian	66
Fries'	133	Carved	67
Grassy	135	Engelmann's	67
Hill's	137	Green Mountain	69
Illinois	138	Lake	68
Leafy	132	Northern shore	69
Long-leaf	141	Prototype	68
Northern snail-seed	152	Spiny-spored	66
Oakes'	143	Tuckerman's	69
Ogden's	117	*Ranunculus*	160
Ribbonleaf	130	*aquatilis*	164
Richardson's	150	*cymbalaria*	160
Robbins'	151	*flabellaris*	161
Small	149	*longirostris*	162
Snail-seed	127	*recurvatus*	160
Spotted	148	*repens*	160
Straight-leaved	154	*sceleratus*	163
Vasey's	155	*trichophyllus*	164
White-stemmed	146	*Riccia*	1
Pond lily	103	*Ricciocarpus*	1
Immigrant	104	Riverweed	115
Small-leaved	105	Horn-leaved	115
Yellow	106	*Rorippa aquatica*	165
Pontederia cordata	116	*Rotala ramosior*	80
Potamogeton	117	*Ruppia maritima*	166
alpinus	122	Rush	70
amplifolius	123	Bayonet	70
berchtoldii	125	*Sagittaria*	167
bicupulatus	127	*cuneata*	169
confervoides	128	*filiformis*	171
crispus	129	*graminea*	172
epihydrus	130	*latifolia*	167
foliosus	132	*montevidensis*	174
friesii	133	*rigida*	175
gemmiparus	134	*teres*	176
gramineus	135	Sago false pondweed	185
hillii	137	*Schoenoplectus subterminalis*	177
illinoensis	138	Sea-wrack	207
natans	140	Sedges	1
nodosus	141	Smartweed	114
oakesianus	143	Scarlet	114
obtusifolius	144	Water	114
ogdenii	117	*Sparganium*	178
perfoliatus	145	*americanum*	178
praelongus	146	*androcladum*	178
pulcher	148	*angustifolium*	179
pusillus	149	*emersum*	178
richardsonii	150	*eurycarpum*	178
robbinsii	151	*fluctuans*	180
spirillus	152	*natans*	181
strictifolius	154	Spatterdock	106
vaseyi	155	*Sphagnum*	1
zosteriformis	156	Spikerush	43
Proserpinaca	157	Spikesedge	43
palustris	158	Little-headed	45
pectinata	159	Needle	44
Pygmy water lily	108	Robbins'	46
Pygmy-weed	36	*Spirodela polyrhiza*	182

St. John's-wort	63
Northern	63
Stuckenia	183
×*fennica*	184
filiformis	184
pectinata	185
vaginata	184
Subularia aquatica	186
Tape-grass	201
Threadfoot	115
Thread-leaf false pondweed	184
Toothcup	80
Trapa natans	187
Typha	1
Utricularia	188
cornuta	190
geminiscapa	191
gibba	192
inflata	193
intermedia	194
macrorhiza	200
minor	196
purpurea	197
radiata	198
resupinata	199
vulgaris	200
Vallisneria americana	201
Water awlwort	186
Water bulrush	177
Water hyacinth	38
Common	38
Water lily	107
Dwarf	108
Fragrant	109
Water lobelia	78
Water lotus	102
Water marigold	22
Water purslane	80
Water starwort	27
Autumn	29
Greater	30
Pond	32
Two-headed	30
Vernal	31
Water star-grass	56
Water violet	60
Watercress	165
Waternymph	95
Canadian	97
Brittle	101
Northern	97
Slender	99
Southern	100
Watershield	24
Waterweed	47
Brazilian	37
Common	48
Free-flowered	49
South American	37
Waterwheel plant	20
Waterwort	39
American	40
Asian	42
Long-stem	42
Small	41
Water-celery	201
Water-chestnut	187
Water-clover	82
European	82
Water-meal	203
Brazilian	204
Columbian	205
Northern	204
Water-milfoil	83
Alternate-flowered	85
Eurasian	91
Farwell's	86
Low	89
Northern	90
Slender	93
Variable-leaved	87
Whorled	94
Water-primrose	80
Common	80
Water-shamrock	82
Water-thyme	61
Wigeon-grass	166
Wild rice	1
Wolffia	203
borealis	204
brasiliensis	204
columbiana	205
Yellow cress	165
Lake	165
Zannichellia palustris	206
Zizania	1
Zostera marina	207

ABOUT THE AUTHOR

Donald J. Padgett, Professor of Biological Sciences, holds a B.S. degree in biology from Susquehanna University, Selinsgrove, Pennsylvania, and an M.S. and Ph.D. in plant biology from the University of New Hampshire, Durham. He teaches wetland ecology, invasion ecology, and general biology at Bridgewater State University, Bridgewater, Massachusetts. His research focuses on the systematics of aquatic and wetland plants. He is also the author of *Aquatic Plants of Massachusetts*, *Aquatic Flora of Connecticut & Rhode Island*, and *Wetland Plants of New England: A Guide to Trees, Shrubs, and Lianas*.

Made in the USA
Columbia, SC
06 August 2024